Science & Technology of
PHOTOVOLTAICS

Second Edition

Science & Technology of PHOTOVOLTAICS

Second Edition

P. Jayarama Reddy
Former Professor of Physics & Vice Chancellor,
Sri Venkateswara University,
Tirupathi

BSP BS Publications

CRC Press
Taylor & Francis Group
A BALKEMA BOOK

Distributed in India, Pakistan, Nepal, Myanmar (Burma), Bhutan, Bangladesh and Sri Lanka by **BS Publications**

Distributed in the rest of the world by
CRC Press/Balkema, Taylor & Francis Group
Schipholweg 107C
2300 AK Leiden, The Netherlands

First published in India by
BS Publications, 2010

ISBN 13: 978-0-415-57363-4

Printed in India by
Sanat Printers, Kundli

Published by
BS Publications
4-4-309, Giriraj Lane, Sultan Bazar, Hyderabad - 500 095 (A.P.) India.

Dedicated To

SAI

The Eternal Source of Energy

Harin S. Ullal, ***Ph.D.***
National Center for Photovoltaics
National Renewable Energy Laboratory
Golden, Colorado, USA

Foreword to Second Edition

This is an excellent book on the science and technology of solar photovoltaics. Great thought has been given in presenting each chapter in a concise and precise manner. It should make a good textbook for physics and engineering students at both the undergraduate and graduate level. Even for a new entrant to the field, it should be a good book on PV technology. My very best wishes to Prof. Reddy for all his commitment, dedication and hard work to put this book together on solar photovoltaic technology.

Harin S. Ullal

Prof. P. RAMA RAO
Former Director, Defence Metallurgical Research Laboratories
Former Secretary, Ministry of Science & Technology, Govt. of India
Former Chairman, Atomic Energy Regulatory Authority
Former Vice Chancellor, Central University of Hyderabad

Foreword to First Edition

Energy is fundamental to human development and to prosperity of nations. The ***Human Development Index*** coined by UNDP is correlated to per capita electricity consumption. The disparity between our country and United States, a superpower, is manifest in the differential in per capita consumption of electricity in India (about 390kwh) and in USA (13000kwh), which is a factor of about 35. India by all accounts is on a trajectory of growing prosperity which can be kept up only through even more rapid growth in energy availability. Clearly then, the country must strive hard to tap every possible source of electricity, thermal, hydroelectric, nuclear and nonconventional. Among the nonconventional energy sources, India has made notable progress in utilising wind energy. However, solar based energy, the other potential and abundantly available resource, is yet to take off in our country. Solar energy is clean and inexhaustible and India has to do all it can to accelerate the pace of development of solar energy. International developments in the area of photovoltaic solar cells and devices have been phenomenal and every effort, such as that presently by Professor P. Jayarama Reddy, in documenting global developments is a much needed contribution and should be most welcome.

Professor Jayarama Reddy is an accomplished teacher and a scientist who has been assiduously working in this field for a number of years. He has taken enormous trouble, true to the tradition of a dedicated teacher, to present in this book the state-of-art relating to several aspects of PV Technology. Following a systematic treatment of the physics of solar cells and the basics of solar cell technologies, Professor Reddy goes on to describe a range of engineering aspects relevant to extracting and utilising energy from solar devices.

In as many as 7 chapters, he has put together in substantial detail technical information on solar modules and arrays, solar photovoltaics systems, storage batteries, power conditioning equipment, on features of system design, on how PV systems are integrated in building structures and finally on aspects of installation, operation and maintenance of PV Systems. The last chapter of the book deals with the market for photovoltaics. In having chosen to include technical features pertaining to design, operation, maintenance and marketing of photovoltaic systems, Professor Reddy has imparted to his book on Science and Technology of Photovoltaics a special character. I have no doubt that the book would serve as a guide to the beginners and as a valuable reference to the experienced specialists in the field of photovoltaics. This is a valuable addition to the extensive literature on the subject and, I trust, will be received with due appreciation from all concerned. He has my very best wishes for every success of his commendable effort.

HYDERABAD
5th May, 2004

P. Rama Rao

Preface to Second Edition

The world-wide progress in Solar Photovoltaics has been phenomenal in the last five years. Significant technological advancement is registered in terms of development of different PV cell technologies, increase in the number of manufacturers of PV solar cells and modules, development of manufacturing equipment for large scale production, increased volumes of production and yearly shipments. PV application spectrum has widened particularly in the field of building-integrated systems and installation of large grid connected power plants on commercial scale. Extensive studies on associated environmental aspects, estimation of energy payback times, cost Vs production analysis and strategies for market development have been carried out. The quantity and the quality of presentations at the international meets in USA, Europe and other countries in recent years reflect the tremendous development of the PV industry, the commercial status and the role of PV solar in global energy sector. The author felt it desirable to revise the book including, briefly, the significant achievements. As a result, many chapters are upgraded, and the ones on Solar cell technologies, Power conditioning equipment and Market development are almost rewritten. A new chapter on 'Energy Analysis and environmental Issues' is added due to its relevance.

This revision has been made possible by the willing support of many active researchers in this area, particularly Dr.H.S.Ullal, NREL, Golden,CO.(USA), Professor N.M.Pearsell and Dr.R.W.Miles, Northumbria University, New Castle,UK, and Dr. Bernhard Dimmler, CEO/MD, Wuerth-Solar, Germany, and Prof. K.T.Ramakrishna Reddy, Sri Venkateswara

University, who have supplied the necessary material. The author expresses his appreciation and profusely thanks them for their valuable help. The support and contributions of Prof. K.S.N.Raju, former Defence Scientist and Professor at Vasavi college of Engineering, Osmania University, Photovoltaic Instrumentation and Power Electronics Engineer and currently an active Solar Energy Consultant, deserve special mention, appreciation and thanks.

A good amount of material has been drawn from many research publications, journals, reports and Company websites. These are fully acknowledged in the 'References' and other places in the Book. The author sincerely thanks these authors of the research papers, publishers of the journals, and several Companies. Due to limitation of space and accessibility of the research material, the author could not include many more publications related to PV Solar in the Book.

The support extended by the management of B.S. Publications – Mr. Nikil Shah and Mr. Anil Shah – and the technical staff, particularly Naresh and Kamesh, in the preparation of the Second edition is appreciated.

P. Jayarama Reddy

Preface to First Edition

The development of solar energy has received special attention due to its abundant availability, its unique suitability for certain vital applications and its potential for replacing conventional fossil fuel sources of energy. In the last few decades, solar energy technologies have evolved from laboratory scale prototypes to commercially viable power plants. This trend, in the last one decade, has attracted heavy investments into the industry in Europe, Japan and USA. Consequently, the production volumes have significantly increased resulting in tremendous fall in prices. Building integrated photovoltaics and large-scale grid connected systems are increasingly being installed in the last one decade. Moreover, its environment-friendly characteristic enables it to fit into the sustainable energy strategy too.

The purpose of this book is to present briefly the present state of knowledge relating to several aspects of solar photovoltaic technologies. Starting with physics of solar cells, the book covers different solar cell technologies, the solar module fabrication, the three types of solar photovoltaic systems and their features, the storage batteries, the power conditioning equipment, the PV system sizing and design, the building integrated photovoltaics, the system installation, operation and maintenance guidelines and the present market for photovoltaics with future projections for the next two decades . The economic analysis is also presented in the last chapter. The solar data for many important cities in India are given in the Annexure.

The material in this book is drawn from author's experience of teaching and research in 'materials' and 'solar energy' for over 20 years in the Department of Physics of Sri Venkateswara University, and later his exposure to PV industry for over a decade in terms of project formulation, promotion, execution and evaluation and conducting training programs sponsored by Government of India for the personnel working in PV industry.

The book is carefully designed to meet in general, the requirements of a student of engineering or science of an Indian University pursuing a course on 'solar photovoltaics'. It is also intended for the people in the middle rung of PV industry to get a overall picture of the PV technology. While the book provides introduction to the complete spectrum of solar photovoltaic technologies, starting from basics of solar cells to economics of solar photovoltaic power generation and details on certain important aspects, the user is expected to go to specific literature for further details.

The author has drawn heavily in terms of material, figures, photos and tables from different journals, publications, company brochures and other leaflets meant for distribution, and they are adequately acknowledged and referred at the appropriate places. The author expresses his sincere thanks to all these authors, publishers and companies, particularly to James & James (Science Publishers) Ltd., London, UK, who are the publishers of Renewable Energy World; to TATA BP Solar India Ltd., Bangalore; to Astro Power Inc., Newark, DE, USA; to Fowler Solar Electric Inc., MD,USA; to Sandia Laboratories, Albuquerque, NM,USA; to Dr. H.W. Schock and Dr. F. Pfisterer, Insititute of Physical Electronics, University of Stuttgart, Germany; to Dr. N.M. Pearsel, Northumbria Photovoltaics Applications Centre, UK; to Arvind Microelectronics Pvt Ltd, Hyderabad and to websites, specially,

www.acre.murdoch.edu.au/refiles/pv.text.html
and *www.sandia.gov/pv/sysd/ResidenceSize.html*

The author gratefully acknowledges the sincere help and contributions from Prof. K.S.N.Raju of Arvind Microelectronics Pvt Ltd., and a former Defence Scientist and Professor at Vasavi College of Engineering, Dr. K.T. Ramakrishna Reddy, Sri Venkateswara University, Tirupati, Prof. S.V. Suryanarayana, Retd. Professor of Physics, Osmania University, and Prof. S. Mohan, Indian Institute of Science, Bangalore in the preparation of the manuscript and to bringing it to the present form.

The author expresses his gratitude and special thanks to the distinguished Materials Scientist, Prof. P. Rama Rao, former Director, Defence Metallurgical Research Laboratories and former Secretary, Ministry of Science & Technology, Govt. of India for expressing his considerate views on the book in the foreword.

The effort and cooperation of my young friends, Mr. R. Ravichander in obtaining a beautiful computerized copy of the manuscript and Mr. P. Ravi Kiran for artistically designing the book are acknowledged. The cooperation and support extended by Mr. Nikhil Shah, M/s B.S. Publications is deeply acknowledged.

Finally, the author expresses his deep appreciation to the members of his family,– wife, daughters and sons-in-law – and well wishers for their constant support and encouragement during the preparation of the book.

P. Jayarama Reddy

Contents

Chapter 3

Solar Modules & Arrays

Chapter 7
Design, Installation and Operation of a Solar PV System

Chapter 8
Building Integrated Photovoltaics

LIST OF COLOR PLATES

About the Author

P. Jayarama Reddy is a former Professor of Physics, Sri Venkateswara University, Tirupati, India. He worked as a Research Associate in the Department of Materials and Engineering, Cornell University, Visiting fellow in Imperial college of Science & Technology, London, Visiting fellow at Charles University, Prague under Exchange Programme, and Guest Professor, Institute of Physical Electronics, University of Stuttgart. He was a Member of Vice Chancellors' delegation to P.R.China. He worked as Director, Sri Sathya Sai Institute of Higher Learning (Deemed University), Prasanthi Nilayam. He had more than 30 years of teaching and administrative experience.

He initiated solar energy research at S.V.University, which attracted special assistance for research from University Grants Commission and Ministry of Non-conventional Energy Sources, Government of India.

He directed several research projects sponsored by CSIR, UGC, DST and DNES/MNES related to materials and solid thin films. He published around 250 research papers along with his students in the areas of anharmonicity of crystals, semiconducting thin films, and thin film solar cells. He traveled extensively, presented papers and delivered invited lectures in International Conferences and chaired technical sessions.

For more than a decade now, he is a nominee Director on the Boards of several Renewable energy companies – solar module fabrication, power generation from biomass, and power generation from solid waste.

His interests include Environmental Studies and published a book, 'Principles of Environmental studies' and, another 'Pollution and Global Warming' is in the press.

About the Authors

CHAPTER ONE

FUNDAMENTALS OF SOLAR CELLS

1

FUNDAMENTALS OF SOLAR CELLS

Solar cells are basic building blocks for the fabrication of photovoltaic modules which are the main component in the solar photovoltaic power generation systems. A clear understanding of the science of the solar cells, the fabrication, the principle of operation, the electrical behavior and the efficiency, are essential to go into the technology aspects of solar photovoltaics. These aspects are discussed in detail in this chapter.

1.1 PHYSICS OF SEMICONDUCTORS

A solid can be a piece of copper or glass or wood; but distinct differences exist among the three due to the arrangement of atoms, the basic constituents in a solid.

In copper, there is a regular pattern of arrangement of atoms, which can be confirmed using x-radiation. Such a solid is called 'crystalline solid', wherein there is a long- range order of arrangement of atoms. In glass and wood, the situation is different.

The atoms in a crystal are held in position by 'bonds'. There are five idealized types of bonding: ionic, covalent, metallic, hydrogen and molecular, and each crystal is characterized by one of these types of inter atomic bonding. However, in some cases, many bonds are formed by two or more of these idealized forms. Of these, the metallic and covalent bonds are relevant for us.

Solids with metallic bonding are excellent electrical conductors because of 'free' electrons (the outermost electrons in the atoms). Metals are mechanically tough and ductile. The solids with covalent bond are generally insulators electrically. They are extremely hard and have high melting points. The elements in the IV Group of the Periodic Table are characterized by covalent bonding.

Between these two categories of good conductors and insulators, there is an important group of solids called 'semiconductors'. These do not have 'free' electrons' to exhibit electrical conduction at low temperatures; but at room temperature, they show a modest electrical conductivity, which increases with temperature increase.

There are many semiconductor crystals, but silicon and germanium belonging to IV group of the periodic table are very important ones. The third group elements and fifth group elements chemically combine to form compounds which are semiconductors. Examples are GaAs, InAs, GaP, InP, InSb and so on. Similarly, group II elements combine with group VI elements forming compounds such as CdS, ZnS, CdTe which are semiconductors.

The electrical conductivity of semiconductor materials generally range from 1/10000 to 10000 mho/cm, which is far less than that of metals. The conductivity behavior of semiconductors may be explained with a simple model. The two-dimensional lattice structure of a typical semiconductor, silicon is shown in Fig.1.1(a). Each silicon atom has four nearest neighbours. It has 14 electrons in its atomic structure and four of them are in the outermost orbit which are weekly-bound to the nucleus. These four electrons are called 'valence electrons' which form 'covalent bonding' with the four nearest-neighbor atoms .ie, each valence electron of a silicon atom is shared by one of its four nearest neighbors.

The valence electrons thus help to bind one atom to the nearest neighbour resulting in these electrons being tightly bound to the nucleus. In a 'chemically pure' or 'intrinsic' silicon which contains no foreign atoms, this ideal situation exists at 0 deg.kelvin, and the crystal behaves as an electrical insulator since no free electrons are available. When the temperature of the crystal is raised or light is allowed to incident on it, the 'bond' gets broken and the valence electron becomes a free electron leaving behind a 'hole' as shown in Fig.1.1(b).

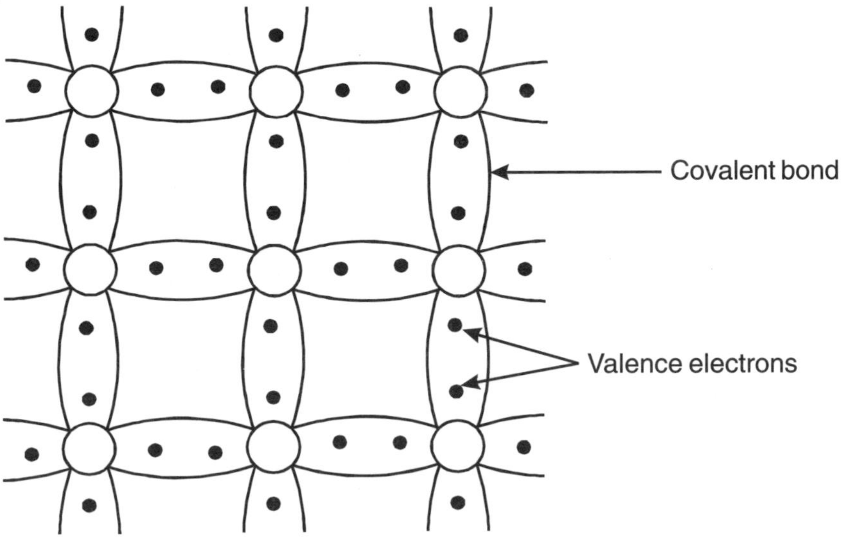

Fig. 1.1(a) *Two-dimensional Silicon Lattice*

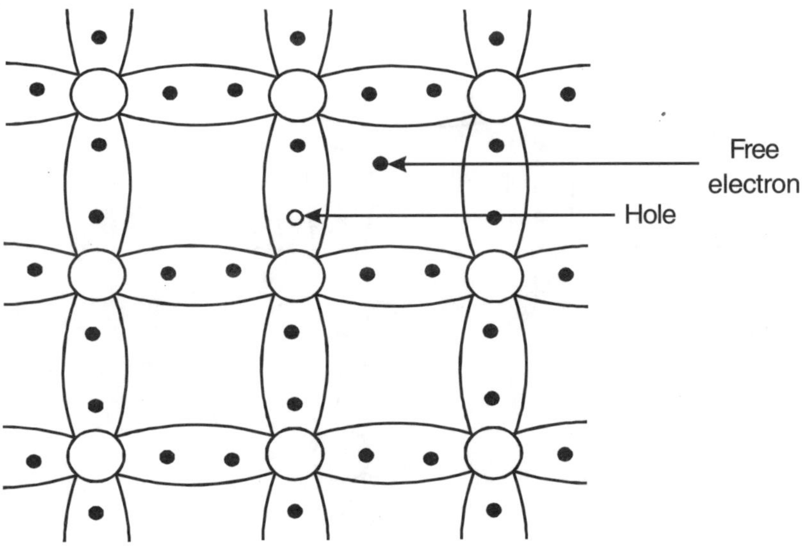

Fig. 1.1(b) *Silicon Lattice with a broken covalent bond*

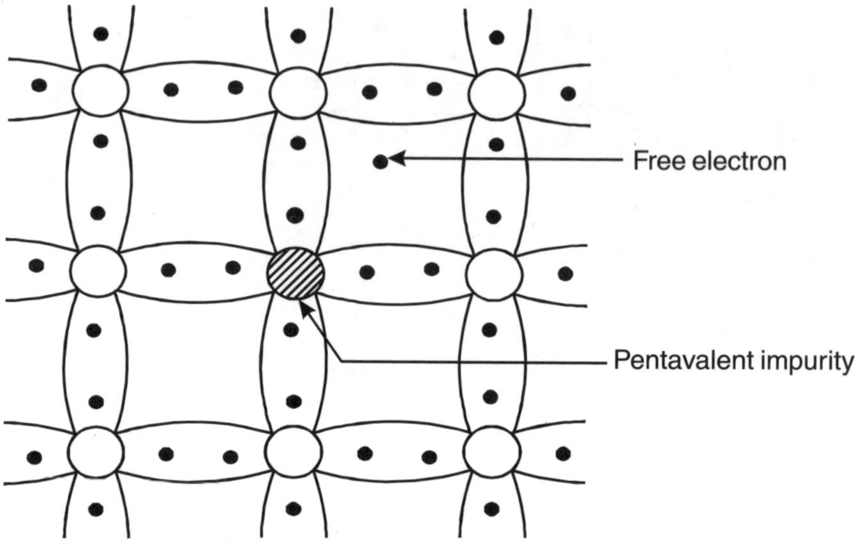

Fig. 1.1(c) *Silicon Lattice with one Si atom displaced by a pentavalent atom*

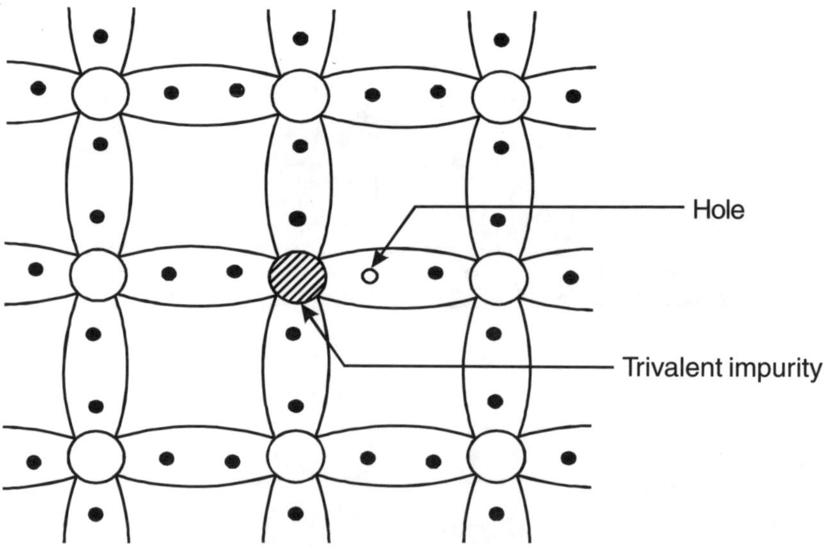

Fig. 1.1(d) *Silicon Lattice with one Si atom displaced by a trivalent atom*

The hole is an incomplete covalent bond. It takes energy equal to 'band energy' E_g, a characteristic parameter of a semiconductor, to separate an electron from the bonding and make it a 'free' electron, also referred to as 'conduction' electron.

For silicon, E_g is equal to 1.1 eV at room temperature. When a hole exists in a bond, it is relatively easy for a valence electron in a neighbouring atom to leave its covalent bond to fill the hole, leaving a hole in its original position. The 'hole' therefore effectively moves opposite to that of electron. The movement of holes, behaving like positive charges, constitute an electric current.

Consider a situation where a silicon atom is replaced by a pentavalent atom like phosphorous (P) as in Fig.1.1(c). P atom has five valence electrons and four of them occupy the covalent bonding and the fifth one nominally held in the bond can move freely as a carrier of current due to thermal excitation. The energy required to detach this electron is about 0.05 eV for silicon, which is very small compared to E_g. By adding pentavalent atoms, the creation of 'extra' free electrons are achieved, the density of these electrons depending on the dopant concentration. This type of doping is called n-type doping and the dopant is called 'donor' type.

On the contrary, a silicon atom is replaced by a trivalent atom like boron (B) which has three valence electrons as shown in Fig.1.1(d). Then bonding with only three nearest neighbors is complete.This atom can capture an electron from a nearby bond, establishing the fourth bond. The captured electron is not available for conduction, but a hole created in the original position of the electron conducts. This is called p-type doping wherein creation of extra holes are achieved, the hole density depending on the dopant concentration. This type of dopant is called 'acceptor' type.

Small amounts of dopants, few parts per million, introduced into the semiconductor structure can significantly increase the number of electrons available for breaking away from their atoms.

The valence electrons are confined to the valence band; but when light energy equal to or greater than E_g is incident, they move into the conduction band, creating an equal number of 'holes' in the valence band. In Fig.1.2, E_c represents the lowest energy that conduction electrons may have and E_v represents the lowest energy that 'holes' may have. E_g is the separation of conduction band edge from valence band edge.

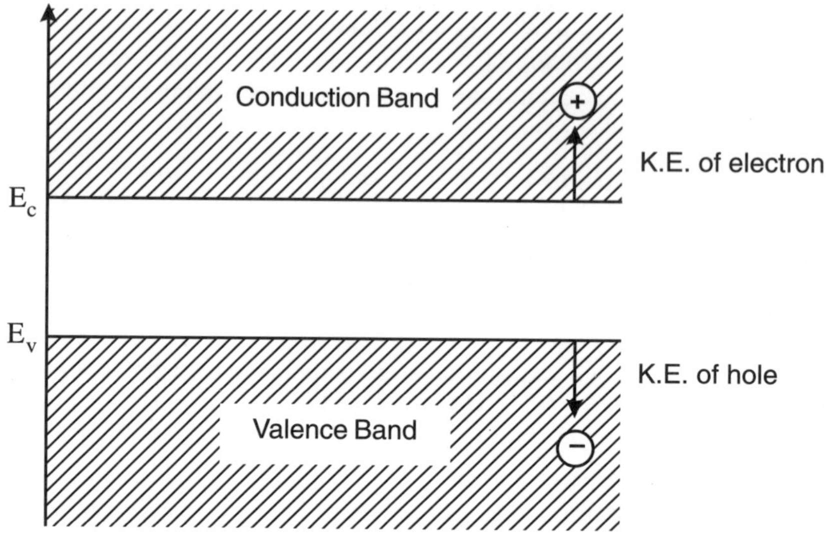

Fig. 1.2 *Energy Band diagram of a Semiconductor,* $E_C - E_V = E_g$

The electron density, ‘n’, and the hole density, ‘p’ are related to the band gap energy, E_g as

$$np = B e^{\left(-\frac{Eg}{kT}\right)} = n_i^2 \qquad \text{.......... (1)}$$

where B is a constant, k is the Boltzmann constant and T is the temperature.

In a pure or intrinsic semiconductor, n and p are equal and represented by n_i, the intrinsic carrier concentration. In silicon, n_i is about $10^{10}cm^{-3}$ at room temperature.

The electrical conductivity may be altered by doping with small amounts of ‘dopants’ such as Group V (donor) or Group III (acceptor) elements of the periodic table. Each donor atom introduced into silicon crystal contributes one extra electron and if the donor density N_d is higher than n_i, then ‘n’ is approximately equal to N_d; and from Equation (1),

$$p = \frac{n_i^2}{n} = \frac{n_i^2}{N_d} \qquad \text{.......... (2)}$$

In the donor-doped semiconductors called n-type semiconductors, n >> p; i.e., electrons are majority charge carriers and holes are minority charge carriers.

For example, a silicon crystal doped with $10^{15}cm^{-3}$ atoms of fifth group element would have 'n' equal to $10^{15}cm^{-3}$ and 'p' equal to 10^5cm^{-3}.

If a third group element is added to silicon, each dopant atom creates one 'hole' and the semiconductor is referred as p-type semiconductor. If N_a is the acceptor density, then

$$p = N_a$$

and

$$n = \frac{n_i^2}{N_a} \qquad \text{.......... (3)}$$

In p-type, p>>n; i.e., holes are majority charge carriers and electrons are minority charge carriers. The electrical conductivity of a doped semiconductor is represented as,

$$\sigma = q\mu_n n + q\mu_p p \qquad \text{.......... (4)}$$

where μ_n and μ_p are the 'electron' and 'hole' mobilities and q is the charge on the carrier. The mobilities of the charge carriers are characteristic of the semiconducting material and are dependent on doping concentration and the temperature of the material.

The nature of semiconductor can be changed by adding 'donors' to a p-type semiconductor and 'acceptor' atoms to an n-type semiconductor. If donor and acceptor concentrations become equal, the semiconductor remains 'intrinsic' because there would be no 'free' charge carriers.

1.2 GENERATION OF CHARGE CARRIERS DUE TO PHOTON ABSORPTION

If a semiconductor of band gap energy, E_g absorbs light energy, hν, greater than E_g, each absorbed photon raises one electron from valence band to conduction band, creating one electron and one hole. Here 'h' is Planck's constant and ν is the frequency of the light wave. The carrier generation rate per unit area of the semiconducting surface is therefore a function of E_g; however, the reflection of incident light at the surface and incomplete collection of charge carriers affect the generation rate. Fig.1.3, indicates that low band gap semiconductors generate more electric current.

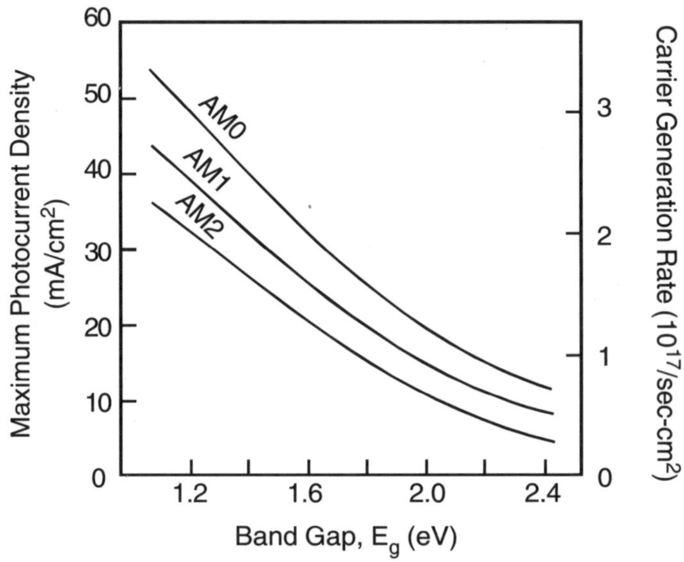

Fig. 1.3 *Band Gap of the semiconductor Vs Maximum Photocurrent Density and Carrier Generation Rate*

Actually, the incident light photons move a distance 'x' from the surface of the semiconductor before being absorbed. The decrease of photon flux from the surface is exponential in nature and the 'flux' at a distance 'x' is

$$P(x) = P(o)\, e^{-\alpha x} \qquad \text{.......... (5)}$$

where α is the 'absorption coefficient'.

The rate of generation of carriers per unit volume is

$$G(x) = -\frac{dP(x)}{dx} = \alpha P(o) e^{-\alpha x} \qquad \text{.......... (6)}$$

The absorption coefficient is a function of photon energy, hν and is zero for hν < E_g. In some semiconductors, α goes to zero abruptly and in others it occurs gradually. The abrupt change is a characteristic of 'direct-gap' semiconductors whereas a gradual change is a characteristic of 'indirect-gap' semiconductors. This behavior is an important consideration for choosing a semiconductor for solar cell fabrication.

Fig.1.4 illustrates the photon absorption rate and maximum cell current as a function of material thickness, for a direct-gap and an indirect-gap semiconductors. It is clear that for silicon, until the thickness is around 100 micrometers and above, all the light photons are not absorbed; whereas in GaAs, a thickness of 1 micrometer is sufficient to absorb all the photons. Evidently, direct-gap semiconductors are the most favored for the fabrication of solar cells due to large saving of the material.

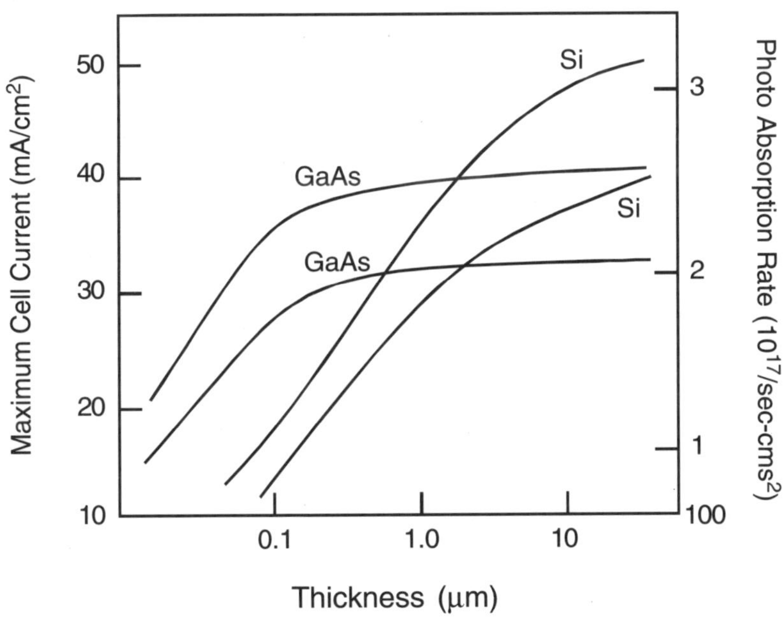

Fig. 1.4 *Thickness Vs Photo Absorption Rate & Maximum Cell Current*

If the carrier concentrations exceed their equilibrium values, the excess carriers 'recombine' and carrier annihilation occurs. The rate of annihilation is

$$\frac{n - n_0}{\tau} = \frac{p - p_0}{\tau} \quad cm^{-3}sec^{-1} \qquad \text{.......... (7)}$$

where n_o and p_o are 'equilibrium' electron and hole densities respectively and τ is the recombination lifetime of the carriers. If τ is high in a semiconductor, it makes a very good solar cell material. In a near-perfect silicon, the recombination lifetime is around a few milliseconds, but in low quality silicon, the lifetime is very low due to the presence of electrically-active point defects, dislocations and even grain boundaries which hasten the recombination process.

Three types of recombination mechanisms operate depending on the nature of semiconductor and the doping density:

(a) *band-to-band recombination, more dominant in direct-gap semiconductors of high doping densities,*

(b) *Auger recombination, dominant in heavily-doped semiconductors,*

(c) *recombination through traps dominant in highly-doped indirect-gap semiconductors.*

These three mechanisms of recombination are schematically shown in Fig.1.5. Generally, only one of these mechanisms predominates over the other two in a semiconductor; however, the total recombination rate is taken as sum of the recombination rates of the three mechanisms.

$$\frac{1}{\tau} = \left(\frac{1}{\tau_b}\right) + \left(\frac{1}{\tau_A}\right) + \left(\frac{1}{\tau_{tr}}\right) \qquad \text{.......... (8)}$$

where τ_b, τ_A, τ_{tr} are the recombination rates due to the three mechanisms respectively. The recombination rate, τ is related to 'diffusion length', L, the distance travelled before the carriers recombine.

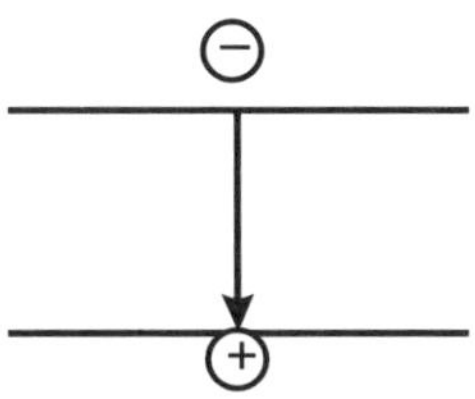

(a) Band-to-band recombination

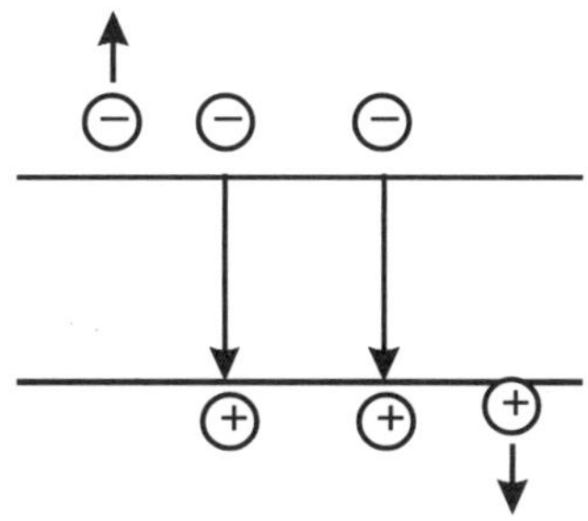

(b) Auger recombination

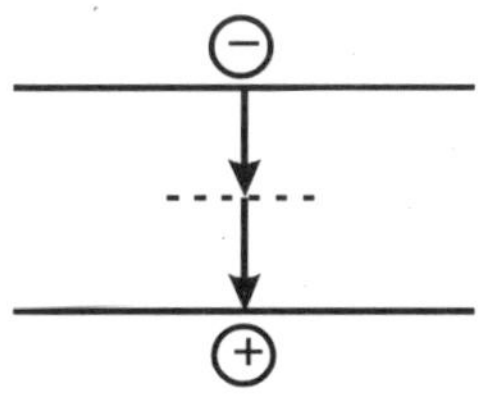

(c) Recombination through traps

***Fig. 1.5** Recombination Mechanisms*

The excess carrier density of the minority carriers decreases exponentially with distance from the plane of the semiconductor where they are generated. The diffusion length, L is

$$L = \left[\frac{kT\mu\tau}{q}\right]^{\frac{1}{2}} \quad \text{or} \quad [\,D\tau\,]^{1/2} \qquad \text{.......... (9)}$$

in terms of 'diffusion coefficient', D equal to $\left[\frac{kT\mu}{q}\right]$

1.3 THEORY OF p-n JUNCTION

Let us examine the operation of a p-n junction. The interface between a p-type region and n-type region of a semiconductor is called a p-n junction. This junction can be created in several ways in a semiconductor. The usual practice is to dope a slice of crystalline semiconductor with donors on one side and acceptors on the other. In Fig. 1.6(a), donor ions are represented by plus signs and electrons they donate by small filled circles. The acceptor ions are represented by minus signs and the holes that are generated by small unfilled circles. Due to higher concentration of electrons on n-side and holes on p-side, the electrons tend to move from n- to p-side and the holes from p- to n-side. The holes which neutralised the acceptor ions near the junction in the p-region have disappeared due to combination with electrons which have diffused across the junction. Similarly, electrons in n-region have combined with holes that crossed the junction from p-region. The unneutralised ions in the neighborhood of the junction result in a charge density. Since this region is depleted by moving charges, it is called 'depletion region' or 'space charge region'. The width 'w' of this region is of the order of a few tenths of a micrometer.

The charge carriers exist only outside this region: to the right, they are predominantly 'holes' and to the left, they are 'electrons'. An electric field and a contact potential difference are created from a non-uniform charge concentration. This contact potential consitutes a potential energy barrier given by (Ref. 1), as shown in Fig. 1.6(b).

$$qV_d = kT \ln\left[\frac{N_d N_a}{n_i^2}\right] \qquad \text{.......... (10)}$$

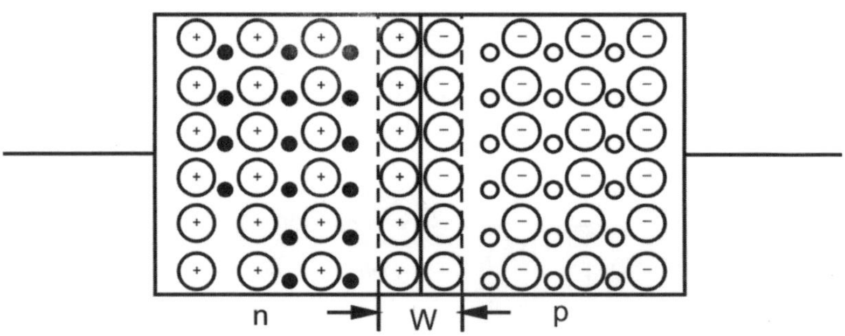

Fig. 1.6 (a) *p-n junction*

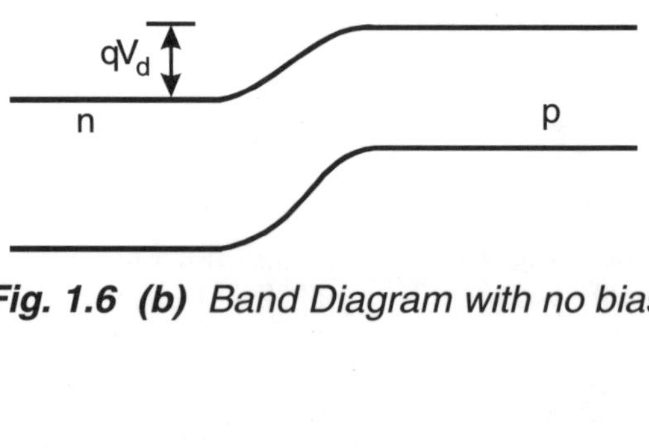

Fig. 1.6 (b) *Band Diagram with no bias*

When a voltage, V is applied across the p-n junction with p-side positive, then the junction is forward-biased and the energy diagram will be as shown in Fig.1.6(c). Here, the external field due to voltage V, opposes the field set up by the contact layer. Consequently, electrons freely move towards p-side and the holes towards n-side, because the potential barrier height is decreased resulting in a net current in the external circuit.

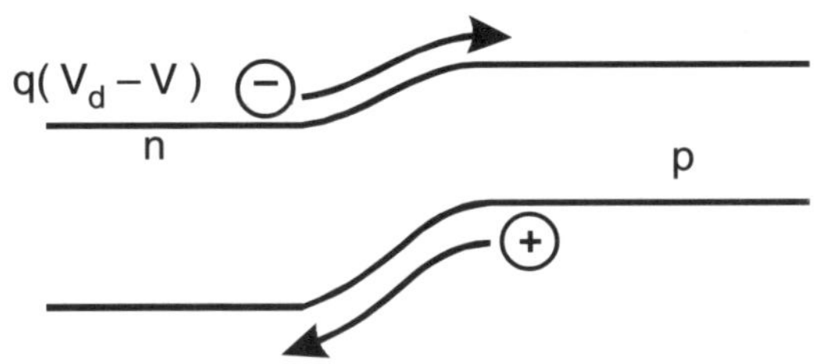

Fig. 1.6 (c) *Band Diagram with forward bias, V*

The electric currents resulting due to the motion of charge carriers may be expressed as

$$\frac{dI_n}{dx} - \frac{n_p - n_{p_o}}{q\tau_n} = 0$$

$$I_n = q\mu_n n_p E + qD_n \frac{dn_p}{dx} \qquad \text{for electrons on p-side} \qquad \text{.......... (11)}$$

and

$$\frac{dI_p}{dx} - \frac{p_n - p_{n_o}}{q\tau_p} = 0$$

$$I_p = q\mu_p p_n E - qD_p \frac{dp_n}{dx} \qquad \text{for holes on n-side} \qquad \text{.......... (12)}$$

Here, I is the current density, D is the diffusion coefficient, and n_p and p_n are respectively the electron concentration on the p-side and hole concentration on the n-side. Assuming that the doping levels on the two sides to be uniform, the equations can be solved for the injected current density, I_{inj}, using boundary conditions.

$$I_{inj} = I_o \left[\exp\left(\frac{qV}{kT} \right) - 1 \right] \qquad \text{.......... (13)}$$

where I_o is known as 'reverse saturation current' of the p-n junction. The current represented by this equation is the dominant one in high quality silicon p-n junctions. But in low quality junctions and in junctions made of high energy gap semiconductors, the 'depletion region-recombination current' which results from the recombination in the depletion region becomes significant.

At high current levels, I_{inj} follow an exponential relationship. The current-voltage (I-V) curves over a current range may be represented by

$$I = I_o \left[\exp\left(\frac{qV}{\gamma kT} \right) - 1 \right] \qquad \text{.......... (14)}$$

where I_o and γ are determined by curve fitting and their magnitudes have to be very low for good performance of the junction.

The important aspect of this expression is that the I-V behaviour of other types of junctions, such as metal-semiconductor and electrolyte-semiconductor and junctions formed by two different semiconductors can also be described by it.

1.4 PRINCIPLE OF OPERATION OF A P-N JUNCTION SOLAR CELL

Solar cell is an electronic device which converts solar energy directly into electrical energy through the photovoltaic effect. It is a typical semiconductor p-n junction device. The principle of operation of a p-n junction solar cell is illustrated in Fig.1.7. When the light falls on the device, the light photons of certain wavelengths are absorbed by the semiconducting material and electrical charge carriers, electrons and holes, are generated. These carriers diffuse to the junction where a strong electric field exists. The electrons and holes are separated by this field and produce an electric current in the external circuit. This current, called photo-current, depends on the incident photon intensity and the nature of semiconductors that constitute the junction device. The merit of a solar cell is that it performs this energy conversion from light to electric energy without producing any noise or pollution and with little maintenance.

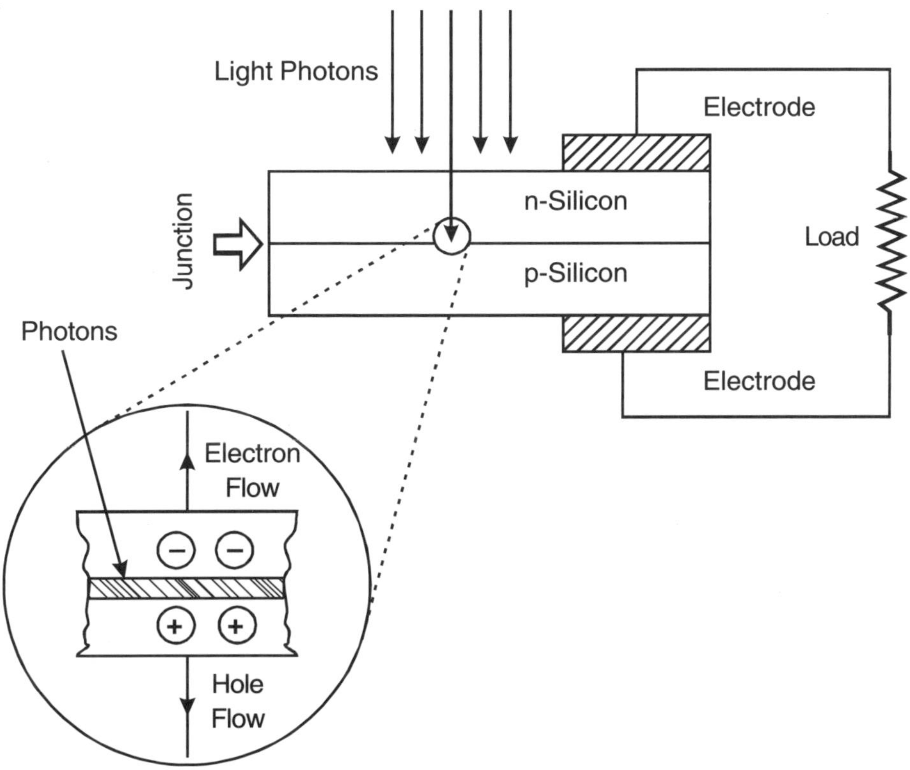

Fig. 1.7 *Photovoltaic Effect in a Solar Cell*

Fig. 1.8 shows the current-voltage curvs, curve 1 represents I-V behaviour with no incident light. Under illumination, a photo-generated current, I_{ph}, flows as reverse diode current, which is linearly dependent on the intensity of incident light.

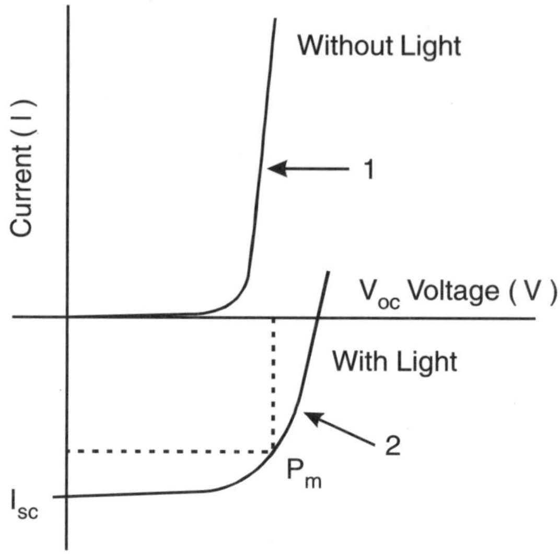

Fig. 1.8 *I-V characteristics of Solar Cell*

The resultant current is

$$I = -I_{ph} + I_o\left[\exp\left(\frac{qV}{\gamma kT}\right) - 1\right] \quad \text{.......... (15)}$$

and is represented by curve 2.

The measurement of I-V characteristics for a solar cell is given in Annexure 1. The output power in the load R_L is $R_L I_R^2$ and indicates the occurrence of energy conversion (Fig. 1.9).

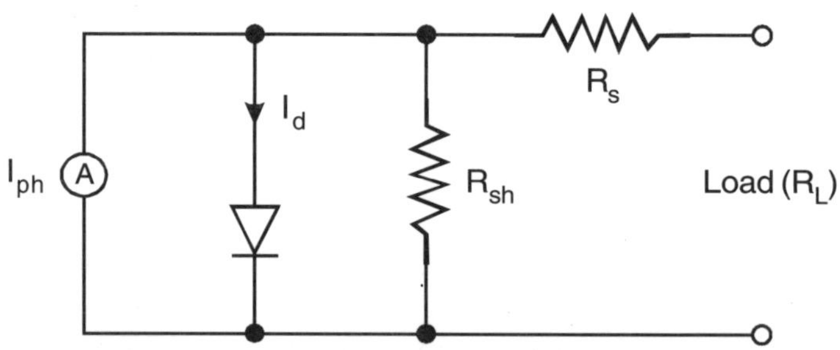

Fig. 1.9 *Equivalent circuit of a Solar Cell*

If the terminals of the junction are shorted, the power output becomes zero; and by adjusting R_L, an optimum value may be found for maximum output power, P_{max}.

The conversion efficiency, η may be defined as

$$\eta = \left[\frac{\text{Output at optimum operating point}}{\text{Energy input to junction}} \times 100 \right] \quad \text{.......... (16)}$$

under certain specified conditions.

Two important parameters - short-circuit current (I_{sc}) and open-circuit voltage (V_{oc}) have to be understood at this point. If the terminals of the junction are shorted, no current flows in the junction. When the junction is illuminated, a current called 'short-circuit current' flows in the junction from p- to n- side or otherwise depending on which side of the junction, the light is incident. This photocurrent is due to three contributions: holes generated in the n-region, electrons in the p-region and electrons and holes in the depletion region before they recombine.

I_{sc} is a function of cell design and is proportional to photon flux. If the load is an open circuit, the corresponding V_{out} is called 'open-circuit voltage, V_{oc}'. I_{sc} and V_{oc} are related to the output power at the 'optimum operating point', P_{max} as

$$P_{max} = V_{oc} \times I_{sc} \times FF \quad \text{.......... (17)}$$

where FF, the fill factor.

$$FF = \frac{V_m \times I_m}{V_{oc} \times I_{sc}} \quad \text{.......... (18)}$$

V_m and I_m are the voltage and current at the optimum operating point. Typically V_m is 75 to 90 percent of V_{oc}; and I_m is 85 to 99 percent of I_{sc} for silicon.

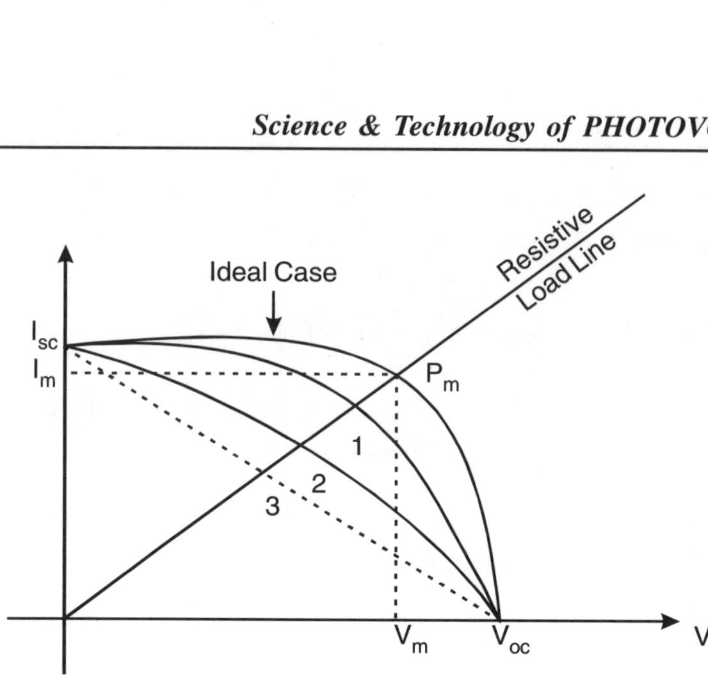

Fig. 1.10 *I-V Curves: 1, 2, 3 as R_s is increasing*

The efficiency expression becomes

$$\eta = \frac{V_{oc} \times I_{sc} \times FF}{\text{Incident Solar Power}} \qquad \text{.......... (19)}$$

In the actual operation of the p-n junction solar cell, the resistive mechanisms and shunt paths come into play significantly (Fig. 1.9); hence, the current-voltage characteristic departs from the ideal one and the fill factor drops (Fig. 1.10).

The efficiency will be high if I_{sc} and V_{oc} are high and so also FF. It means the dark current must be low. The sharp corner in the I-V characteristic is an indicator of high FF.

In practice, the conversion efficiency of a solar cell is measured under Standard Test Conditions (STC), which include (i) solar irradiance intensity of 1000 watts/m^2, (ii) AM 1.5 solar reference spectrum, and (iii) temperature of cell, during measurement, of 25 ^{o}C.

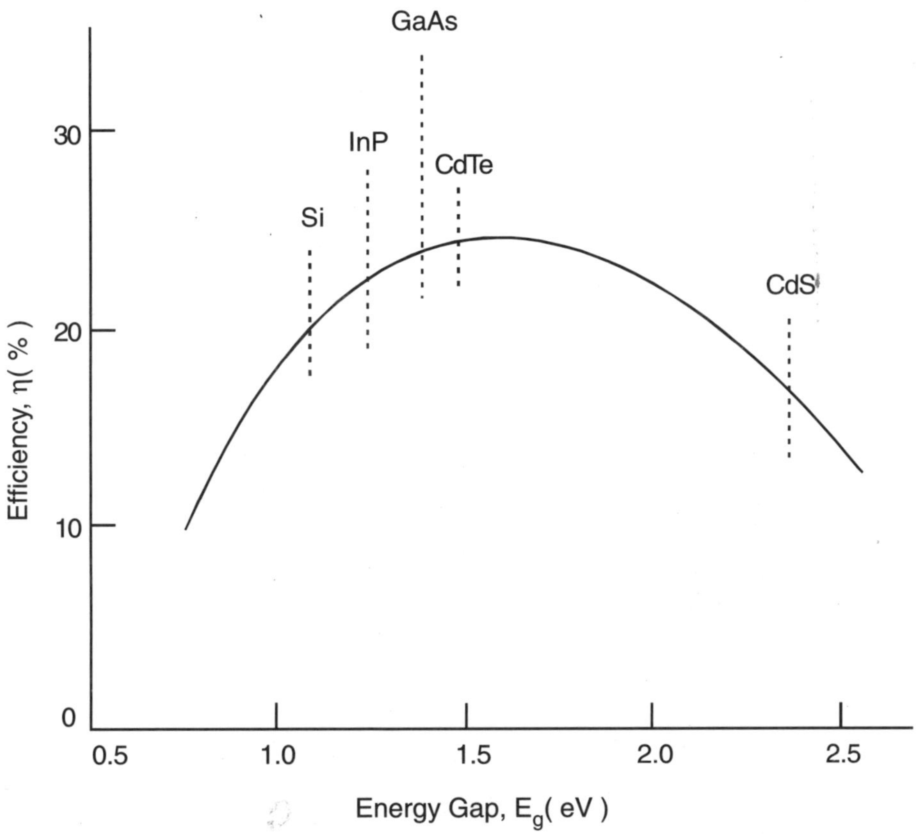

***Fig. 1.11** Estimated Efficiency Vs Energy Gap of the Cell Material*

1.5 RELATION OF V_{OC} AND E_g

The V_{oc} and E_g are related to a first approximation as

$$V_{oc} = \frac{E_g}{q} \qquad \text{.......... (20)}$$

indicating that V_{oc} is directly proportional to the band gap energy.

Earlier in Fig. 1.3, it is seen that the current generation rate decreases with increasing energy gap, which means that the cell efficiency reaches maximum at certain range of values of E_g.

It is interesting to study how the change in energy gap influences the energy conversion of a solar cell. Firstly, a larger E_g reduces the reverse saturation current and increases V_{oc} which in turn tends to increase the efficiency. Secondly, a larger E_g means that a fewer photons can be absorbed because only those photons greater than and equal to E_g are absorbed, which in turn decreases the efficiency. The net result of these two opposing effects are shown in Fig. 1.11. The calculated efficiency has a maximum of about 25% at E_g equal to about 1.5 eV and falls off on either side of this value.

The semiconducting compounds, Cadmimum telluride(CdTe) and Gallium arsenide(GaAs), Indium phosphide(InP) have E_g around this value. In practice, semiconductors with energy band gaps in the range 1.0 to 1.7 eV are used for maximum efficiency devices/solar cells. These include monocrystalline and multicrystalline sillicon(E_g = 1.12eV), copper indium diselenide(E_g = 1.05eV) and amorphous silicon(E_g = 1.7eV).

The silicon-based devices are 'homojunctions' (junctions within the same material), whereas many of the compound semiconductor devices are 'heterojunctions' (junctions between two dissimilar materials). These generally use n-type wide energy band gap layer to form a junction with the p-type absorber layer. This structure is generally preferred.

The theoretical efficiencies are evaluated by assuming an ideal case of a cell fabricated with well-established processing and manufacturing techniques.

The factors which affect the efficiency of solar cells are dealt in the next chapter.

REFERENCES

1. ***Muller,R.S. and Kamins,T:*** *Device Electronics for Integrated Circuits, Wiley-Inter Science, New York (1977)*

BOOKS FOR FURTHER READING

1. ***Larry D.Partin:*** *Solar Cells and Their Applications, Wiley-Inter Science, New York (1995).*
2. ***Richard Bube:*** *Photovoltaic Materials, Imperial College Press, London (1998).*
3. ***M.A. Green:*** *Solar Cells: Operating Principles, Technology and Systems Applications, Prentice-Hall, (1992).*
4. ***J. Nelson:*** *The Physics of Solar Cells, Imperial College Press, London (2003).*

CHAPTER TWO

SOLAR CELL TECHNOLOGIES

2

SOLAR CELL TECHNOLOGIES

2.1 INTRODUCTION

Though the photovoltaic effect was discovered in 1839 by French scientist, Henri Becqueral, a deeper understanding of the scientific principles of the effect was made possible by the scientific investigations of Albert Einstein and of Schottky many years later. The first silicon solar photovoltaic cell was developed at Bell laboratories in 1954 with a conversion efficiency of 6% (Ref.1). Around the same time, scientists from US Air force Aeronautical Research Laboratory reported about 6% efficiency for solar energy conversion using metal - cadmium sulphide junction (Ref.2). The type of cell developed at Bell Laboratories was used to power a space satellite in 1958 for the first time. Since then, solar PV cells have been the source of extremely reliable electric power in all space programmes. The scientists at Bell laboratories also found that the silicon's sensitivity to light could be increased by adding 'impurities' or 'dopants' as observed in most semiconductors.

Silicon is the most abundant element available on earth, and the earliest solar cell was made of crystalline silicon. Today, silicon is used in single-crystalline, polycrystalline and amorphous nature for the fabrication of solar cells. Bulk of the commercial solar cells use single-crystalline silicon.

The polycrystalline and amorphous types of silicon are a result of engineering of silicon material where lattice perfection is sacrificed to achieve significant reduction in the costs of silicon technology for solar cell applications. Many other semiconducting compounds have also been investigated. Solar cells based on Gallium arsenide, Cadmium telluride, and Copper indium diselenide are now commercially available.

2.2 SINGLE CRYSTALLINE SILICON CELLS

The single-crystalline silicon solar cells are the first generation of solar cells and bulk of the PV technology builds on this cell. Crystalline silicon is the extensively researched and understood semiconducting material compared to other solar cell materials. Pure silicon ingots are prepared by single crystal growth on a seed pulled from the silicon melt. The principle of Czochralski process is predominantly used for single crystal growth. In this process, a seed of single crystal silicon comes in contact with the melt of high purity silicon, and as the seed is slowly raised, the atoms of the molten silicon adopt to the pattern of single crystal seed as it cools and solidifies into a single crystal structure (Fig. 2.1). From this single crystalline ingot, wafers are sliced and textured to improve the solar energy absorption. These wafers are then fabricated into p-n junctions by high temperature diffusion of dopants, mainly boron and phosphorous that modify the surface layer composition.

The silicon ingot must have a near-perfect lattice structure and must be of very high purity for the cells to be efficient. The main problem is the loss of good crystalline material as sawdust during cutting process to make wafers. Silicon material has been thoroughly studied; the physical understanding and the device-processing procedures are well established. Silicon solar cells have proven their credibility as source of reliable electric power both on land and in space. Efficiencies in the range of 14 to 17% have been measured for the mono crystalline silicon cells (Ref. 3).

Flat zone crystal growth is another dependable process to obtain single-crystal ingot from polycrystalline silicon rod. A silicon rod is set atop a seed crystal and local heating is carried out by an electromagnetic coil to melt a small zone of the rod. This molten zone starting at the interface between the rod and the seed moves slowly as the coil moves down the length of the rod forming single crystal structure. The merit of this technique is that the crystalline rod becomes single crystal ingot without any impurities and defects.

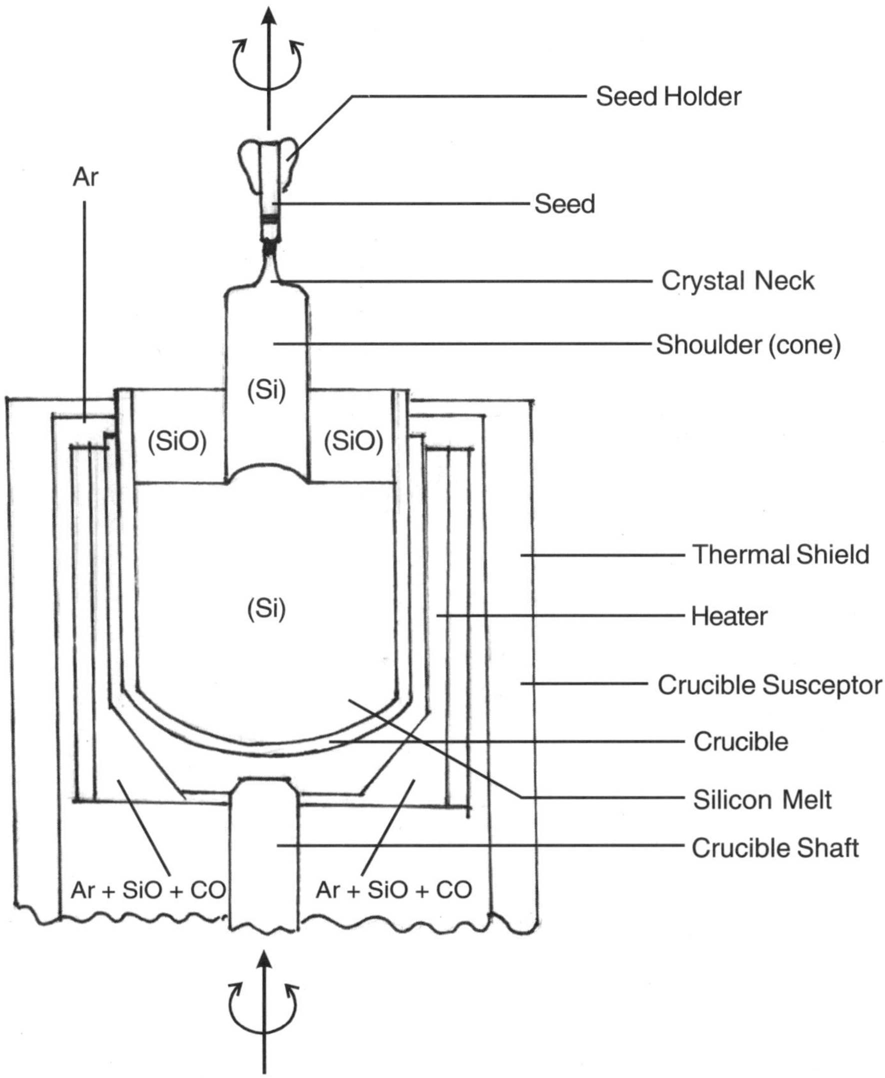

Fig. 2.1 *Czochraski Crystal growing Process*

The highest efficiency verified for the monocrystalline silicon is over 24% (Ref.4, 5). The standard technologies for commercial silicon cells are based on boron-doped p-type (Cz) sustrates having resistivity of 1 to 10 ohm-cm. But the metallization process used is quite different for space and terrestrial applications.

The screen printing technology is used for terrestrial cells whereas photolithography and vacuum deposition are used for the space cells. The screen-printed crystalline silicon solar cells which have dominated production have been on the field for more than two decades(Ref. 6). The modern screen-printed crystalline silicon solar cell is shown in Fig. 2.2 (Ref. 6, 7, 8). Wafers are sawn from a large bowl of silicon grown using Czochralski method. The crystal is boron-doped p-type. Crystalline silicon has an indirect band gap resulting in low absorption coefficient; hence the wafer thickness has to be greater than 200 micrometers to absorb most of the incident solar light. The wafers are mechanically polished to remove damage from sawing, and then chemically polished to remove damage from the mechanical polishing process. The surface is then textured by dipping into a solution of NaOH and isopropyl alchohol to minimize reflection losses and to enhance the optical path lengths in the silicon. A p-n junction is produced by diffusing phosphorus into the wafer to convert the surface to a n-type region. The electrical contact is made to the

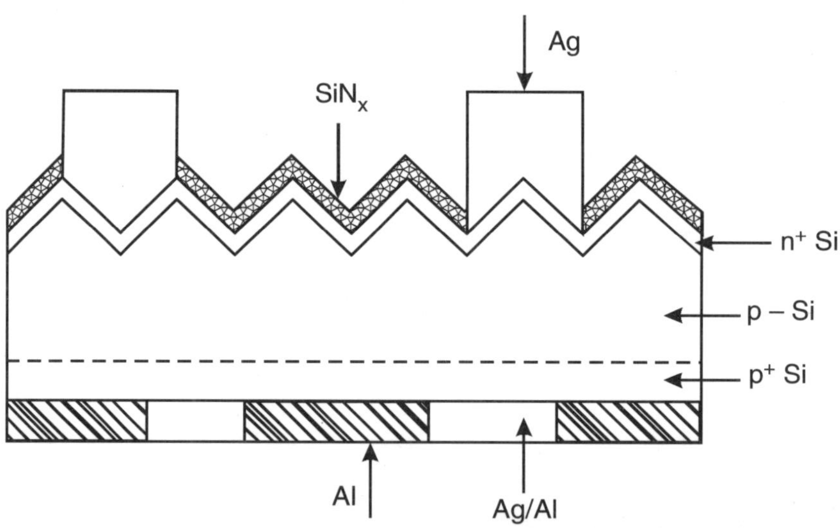

Fig. 2.2 *Screen - printed crystalline silicon solar cell*

n-type region with screen-printed Ag contact fingers that allows light to be transmitted to the depletion region of the p-n junction. The reflection losses are minimized by coating an anti-reflection layer, either titanium dioxide or silicon nitride. The electrical contact to the back p-type surface is done with aluminium paste, which is annealed to introduce a p+ region at the back of the cell to minimize the back-contact resistance and to introduce a back surface field that reflects minority charge carriers back toward the junction (Ref. 9, 10).

Several improvements have been made to the screen-printed metalization process in the last few years. For example, the screen printing process based on firing during/after PECVD SiN_x antireflection layer, appears to enhance the efficiency of industrial solar cells.

Several functions such as front and back contact firing, surface and bulk passivation and Al-formation are combined to make the process more convenient. The screen printing technology is convenient to handle industrially yet efficiency of these cells could not be taken beyond 13 to 15% even with several developments.

Green and co-workers at University of New South Wales(UNSW) have shown that the screen-printed contacts result in reduction of the cell efficiency to about 14% due to high shading losses, nearly 3-times high resistivity compared with pure Ag, a high contact resistance between the grid and silicon, and poor aspect ratio. To overcome this, solar cells wherein the contacts are defined using either photolithography or laser scribing have been developed (Ref. 11, 12).

The Buried Contact c-Si solar cell is shown in Fig. 2.3(a). Mono crystalline Silicon(p-type) wafer with a random pyramid textured front surface to reduce light reflection is the main feature of the cell. The contacts are buried into groves which are machined with a laser. The contact metals, nickle, copper and then silver are deposited using electroless methods. The rest of the surface is covered with a lowly doped emitter which is effectively passivated with SiN_x. Due to low shadowing losses and the good electrical properties of this emitter, the blue response and the overall performance of this cell structure is very high.The higher doping concentration of this n^+ region also reduces the contact resistance at the grid contact. BP Solar has commercialised this technology as its premium 'Saturn' module and has expanded production capacity targeting 80 MW/year by 2006 (Ref. 11). Their Saturn-7 modules demonstrate an 11% increase in DC electrical energy output over the previous modules due to increased cell effeciency and the use of larger cell area, 125mm pseduo square cells(Ref. 13).

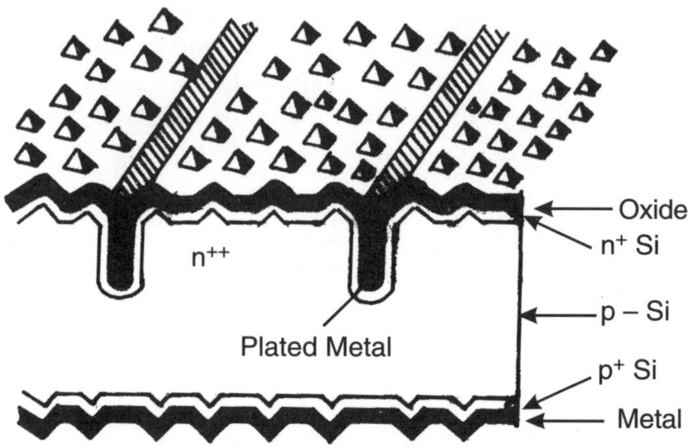

Fig. 2.3(a) *Buried Contact Silicon Solar Cell*

A more advanced design is 'the passivated emitter rear locally diffused' (PERL) solar cell which has shown an efficiency of 24.7% (Ref. 10, 14). In this device, a layer of SiO2 is formed at the back of the device to passivate the back surface and to reflect light back into the cell. Then contact is made by diffusing Al into the Si layer via small windows opened in the oxide layer to produce localized p+ regions (Fig. 2.3b).

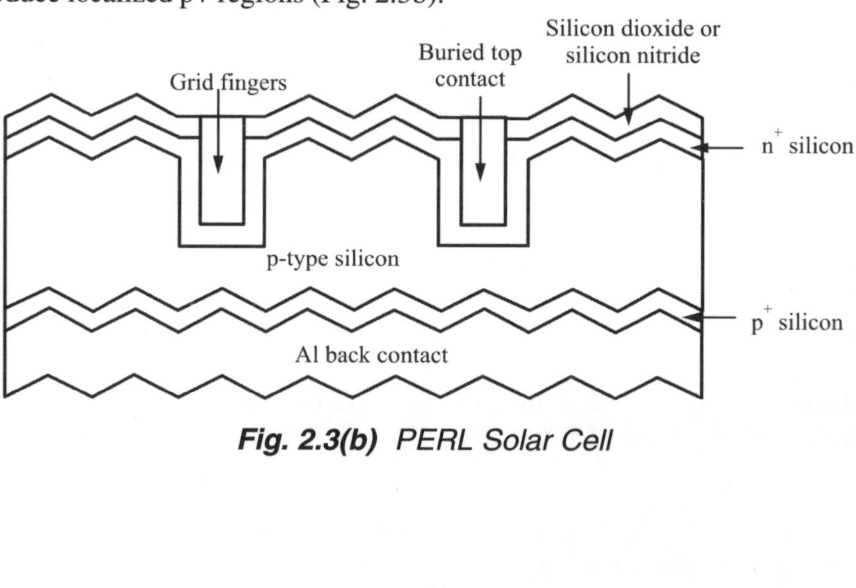

Fig. 2.3(b) *PERL Solar Cell*

The surface is also textured to produce a surface of inverted pyramids which can effectively refract light to high angles of refraction. This aspect permits thinner slices of Si to be used to make cells. This design has resulted in the production of laboratory based cells with efficiencies greater than 24% (Ref. 14,15) and foot-square modules with efficiencies greater than 20% (Ref. 13). It has been found that doping silicon with boron induces a defect that leads to a light-induced degradation of minority carrier life-time in p-silicon, and hence to a drop in efficiency of solar cells by about 1 to 2%. Experiments have shown that if gallium is used to dope the Czochralski grown silicon wafers instead of boron, there is no light-induced degradation in solar cells fabricated with these wafers. This confirms that gallium is a more promising dopant than boron to obtain p-type Si wafers (Ref. 16).

The commercial mono-Si mod ules have efficiencies in the range, 10-16%, and more sophisticated cell designs having higher efficiencies (Ref. 8).

Mono-crystalline Si solar modules had dominated production as well as sales till around 2000, and their market share has reduced to about 40% due to competition from multi (poly) crystalline Si cells.

2.3 POLY (MULTI) CRYSTALLINE SILICON CELLS

Another approach to silicon technology is to prepare polycrystalline silicon block with no lattice perfection. The most popular commercial process is the 'casting process' wherein the molten silicon is poured into rectangular mould and cooled resulting in silicon ingots with large columnar grains growing from the bottom of the container upwards. The ingot is cut into bricks and the low quality regions at the bottom, top and edges of the bricks are removed. Then they are wire-sawed into wafers. The maximum size of the ingot is restricted by the crystallization crucible. Deutsche Solar, one of the biggest manufacturers of mono- and multi-crystalline silicon wafers has a casting unit that gives ingots of size, 270 kg. A 'hybrid' approach to increase the ingot size has been developed by them (Ref.17). In this approach, certain amount of silicon is preheated in the crystallization crucible, referred to as 'pre loaded silicon', before the liquid silicon from the melt crucible is added. The pre loaded silicon melts, assisted by a heater located at the bottom of the crucible on a movable isolation plate. When the entire pre loaded silicon becomes liquid, the bottom heater is switched off and the entire melt solidifies under SOPLIN conditions (Ref. 17) into ingots. By this approach, ingots up to 330 kg, the biggest ever reported, are prepared. In this process, the pre heating temperature, the pouring temperature of the melt, and the amount of energy transferred into the crucible after pouring, decide the complete melting of the pre loaded silicon. The grain size at the bottom of the ingot is similar to that in the pure casting process. It has been found that the grain size in a 310 kg ingot increases with height. In the first few millimeters of the ingot, the average grain size is around 5mm. It increases to around 8mm at 30mm height, and exceeds 10mm at approximately 100 to 130 mm height. Though the low value of grain size at the bottom is reasonably good the bottom layer is removed because the life time of minority carriers is too low here.

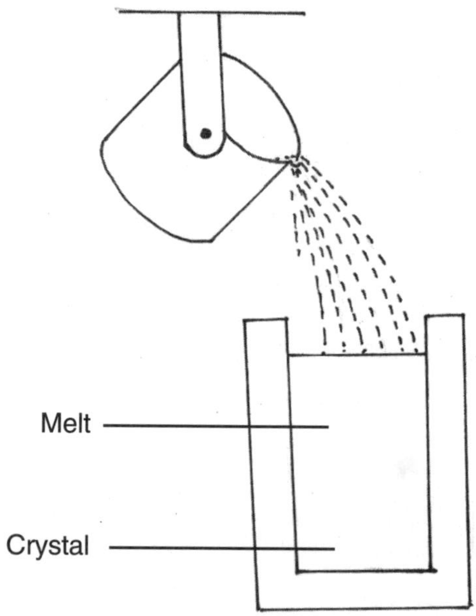

Fig. 2.4 *Principle of Casting Process*

The ingot is then sliced into wafers. This means that the process of cutting and polishing and the waste resulting from this process are much the same as that required for single crystal silicon. Hydrogen is incorporated during device processing which helps in passivating the grain boundaries in the devices formed. This is most conveniently introduced by using PECVD (Plasma Enhanced Chemical Vapour Deposition) to deposit silicon nitride as the top insulating layer (instead of SiO_2) as hydrogen is used in this process. Except this step, the cell processing is similar to that used for mono-crystalline solar cells.

The poly technology has lowered the costs of PV technology because the 'casting' process is relatively cheaper and less sophisticated compared to Czochralski process. The other advantages of this technology are lower capital costs, higher throughput, less sensitivity to the quality of the silicon feedstock, and higher packing density of cells to make a module because of the square or rectangular shape of the cell. The multi-crystalline Si modules generally have efficiencies, 2 to 3% less than those of mono-crystalline Si modules , and cost-wise, these are approximately 80% of that of mono-Si cells to produce. The average efficiency of commercial multi-crystalline Si modules is around 12.6%, not very much different from the average efficiency, 12.8% of commercial mono-crystalline Si cells (Ref.8). This shows that the average multi crystalline silicon modules have caught up with average mono crystalline silicon modules. Willeke (Ref.8) observes that this is due to process improvements relating to crystal growth and hydrogen passivation for multi crystalline silicon, while poorly performing processes are still used for mono crystalline silicon. The manufacturers are 'not making good use of high efficiency potential' of mono silicon.

The market share of multi-crystalline Si modules has been on the increase in the last five years. The building architects feel that appearance of these modules are more appealing to use in the buildings for power generation than the mono-Si modules.

There have been extensive developments in the manufacturing of poly silicon. One method is to deposit a polycrystalline Si film on a suitable substrate by sputtering or other evaporation technique which has the advantage of low material consumption, low energy intensive, no cutting and polishing and scalable and hence to go in for mass production. In this type of cell, the film thickness is small compared to the size of the polycrystalline grains and the junction is inside the film.

Another process is to prepare ribbon and sheet silicon where sawing of the ingot and the associated wastage can be avoided and this has the potential to be a low cost technology(Ref. 18). Here, a silicon sheet is drawn from a bath of molten silicon in a slow and continuous process. This technology is almost on par with the single-crystal option in terms of efficiency but is much lower cost-wise.

There are two technologies to prepare multi-silicon in sheet or ribbon form by drawing, namely, EFG (Edge-defined Film-fed Growth) and SR (String Ribbon) technologies.

The EFG process works by utilizing the capillary force of molten silicon. A graphite shaping-die is positioned in the silicon melt, the essential element being a gap through which molten silicon is driven upwards by capillary action.(Ref.19). A seed frame is approached to the top of the die, until a meniscus is formed between the frame and the graphite die. The seed frame is then drawn upwards at a rate of about 2 cms/minute, and the molten silicon solidifies as a growing sheet beneath the frame. If the drawing speeds are high, high dislocation densities (lattices defects) are formed which reduce the electrical conductivity.

In the improved 'Edge-defined Film-fed Growth(EFG), silicon is shaped into a hollow octogonal tube during the growth itself using a special die, with flat poly crystalline silicon sheets forming the eight vertical sides. Each tube is 5.3 meters in length and each side is about 10 cm wide and about 300 micrometer thick. Wafers of 10 cm x 10 cm are cut from the sheets with Nd:YAG laser which significantly minimises the wastage due to cutting. Wafers of different

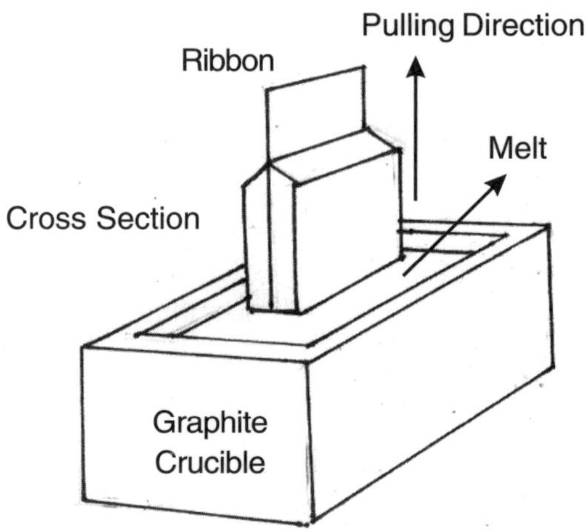

Fig. 2.5(a) *Principle of EFG Process*

lengths are also cut. This enabled to fabricate large area solar cells with efficiencies above 14% (Ref.3).This is a most economical and environment-friendly wafer production process. Further, the silicon losses in the die or crucible are less than 0.5%. AES Americas setup an automated solar cell line for the EFG wafers of 10 cm × 10 cm and 10 cm × 15 cm sizes.

Smart Solar Fab (of Schott Solar, Germany), the proponent of the process produce octagonal tubes with drawing heights of 7 meters, and an edge length of 125 mm with a wall thickness of 300 mm. With laser cutting, 5-inch silicon wafers are made for cell manufacturing (Ref.19).

The production of EFG wafers demands very precise control of temperature. The silicon melt must be hot enough to avoid premature crystallization at the top edge of the graphite die, but at the sametime, cool enough to solidify the seed. Studies by Patric Geiger (Ref.20) specify a necessary temperature stabilization of ± 1°C, a difficult task to achieve.

Wafers move through the line (80 m long) continuously within a time period of about 2 hours. Wafer cleaning, phosphorous diffusion, oxide removal, deposition of antireflection coating, back contact formation, front contact formation, cell testing and inspection are the process steps sequentially involved. The batch cell efficiency averages are around 14%.The cells are used for powering satellites also because of their reliability and performance. The cylinder technology has been successfully demonstrated, that it is not only cost efficient but has the ability to overcome certain limitations of EPG growth.

The symmetry considerations in the cylinder reduce thermo-elastic stresses and by rotating the cylinder during growth, the non-uniformity in the temperatures can be overcome. Cylinders of 50 cm diameter, 1.2 mt-long with wall thickness down to 100 micrometer have been grown with steady-state growth at about 2.5 cm/minute (Ref.21). The principle of the growth process is shown in Fig. 2.5a. This technology is expected to significantly reduce the costs in wafer and cell production.

The String Ribbon (SR) technology was developed by Evergreen Solar, USA. In this process, two graphite strings spaced about three inches apart are pulled through a bath of molten silicon in an argon atmosphere so that a film of liquid silicon stretches between the strings (Ref.22). This solid – melt interface acts as a seed (Fig. 2.5b). The approximate 7 mm high meniscus ensures a stable drawing process, which continues at a rate, 1 to 2 cm/minute. The thickness of this polycrystalline Si sheet is dependent on the speed with which the strings are pulled through the melt. Evergreen has improved this process and now produces two ribbons from a single furnace (Fig. 2.5c), doubling the capacity. The cell process has a reduced number of steps. The cells are rectangular, three inches by six inches, thus ensuring high utilization of the surface area of the module (Ref. 20, 23). SR technology is not as temperature sensitive as EFG process and demands a temperature stability of ± 10° C.

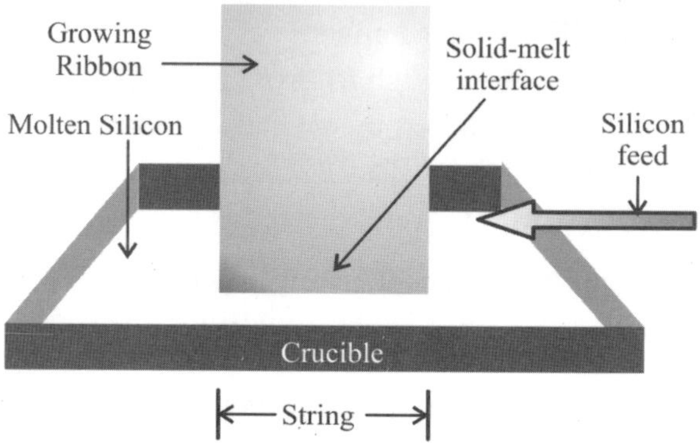

Fig. 2.5(b) *String-Ribbon process from Evergreen: The molten silicon is drawn between two parallel wires (strings) and solidfies into a continuously growing ribbon.*

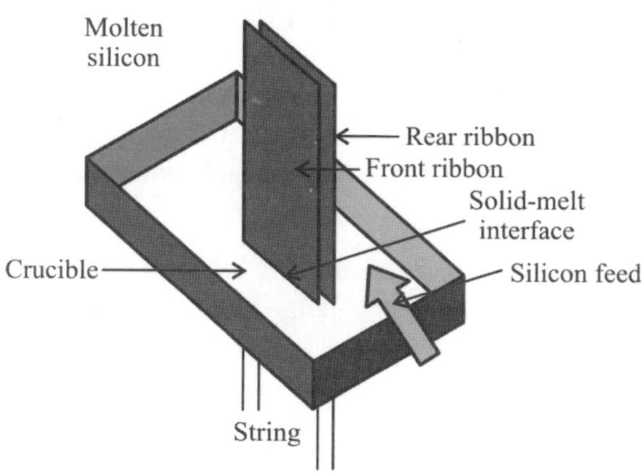

Fig. 2.5(c) *The extremely thin steel wire is threaded through the sawing machine from the supply spool. The individual cuts are only 0.18 mm apart, with the result that the circulating wire appears to form an unbroken surface.*

However, the advantage in both the processes is the saving of considerable amount of silicon, nearly 30% less material than with the sawn wafers. The main disadvantage of both the processes is low efficiency of the cells due to high defect concentration in sheet silicon. Molten silicon is highly reactive and attacks the material with which it comes into contact during the drawing process. That is why, EFG wafers display high levels of carbon impurity attributable to graphite die and also a wavy surface. In the SR process, not only the contamination due to strings but the different grain densities of the silicon sheet affect the efficiency. Current research efforts at both Schott Solar and Evergreen are aimed at improving the electrical performance of the cells to compete with mono-Si; and in the case of EFG, to bring down thickness so that silicon consumption per Wp may be cut from the present 10 gms/Wp to 4 gms/Wp by 2020.

Poly silicon wafers are treated with elements such as hydrogen, lithium, aluminium, arsenic and phosphorous to prevent recombination of charge carriers at the grain boundaries and at other active defects. The best bulk multi crystalline silicon cells are reported at 18.6% efficiency (Ref.4).

The screen printing process is well suited to multi-crystalline substrates also; and 12.5 × 12.5 cm multi-crystalline Si cells with single Si_3N_4- AR coating have demonstrated efficiencies of 15.7%. Recent application of Rapid Thermal Processing (RTP) in the place of conventional furnace processing is another attractive feature for solar cell fabrication because it reduces process time, cross contamination and thermal mass. The RTP technology is successfully applied for junction formation, for the growth of a passivating oxide, annealing of Si/dielectric interface and Si/metal contact firing. There have been many other attempts to improve the efficiency and to reduce the cost of the crystalline silicon solar cell technologies.

The development of crystalline silicon solar cell technology from 1954 to 2004 and the goals for future developments are stated by Willeke (Ref. 8). In this paper the details of the road map prepared by European Photovoltaic Industry Association (EPIA), and the future technical goals for the crystalline silicon are mentioned (Table 2.1).

Table 2.1: Crystalline-Si milestones in the EPIA road map.

Solar grade silicon	**2005**	**Operational plant** of 1000 tons/year, increasing to meet the demand of the PV industry
Wafer	**2010**	Materials consumption for Cz-Si and (2020) mc-Si from 16g/Wp to10g/Wp **(continuing to 8g/Wp)** **Ribbons** from 10g/Wp to 6g/Wp**(to5g/Wp)** **Wafer thickness** from 300ì to 180ìm **(to 100ìm)** **Kerf loss** from 250ìm to160ìm **(to150ìm)**
Solar cells	**2010**	**C- Cz efficiency** from 16.5 to20% (2020) **(continuing to 22%) mc efficiency** from 14.5 to 18% **(to 20%) Ribbon efficiency** from 14 to 17% **(19%)**
Modules	**2010**	**Life time expectancy 35 years,** e.g.longer life time encapsulation materials **Interconnect technology,** e.g. back-contact solar cells
		Here, C-Cz: crystalline Czochralski Si; mc: multi crystalline

2.4 AMORPHOUS SILICON CELLS

Unlike mono and multi-crystalline silicon cells, these are prepared as thin films of silicon. These films have amorphous structure, a structure having no long- or short-range lattice order. These films are deposited onto large, flat substrates of glass or stainless steel or polymer through a glow-discharge, sputtering or any vacuum evaporation method. Thin film technologies brought revolution into solar cell manufacturing because they are cost effective as far as materials are concerned. Amorphous silicon (a-Si) films, however, contain a large number of structural defects.

Hydrogenated amorphous silicon (a-Si: H) films are prepared using plasma-enhanced chemical vapour deposition (PECVD) method (Ref. 24) or hot wire CVD (Ref. 25). These films proved to be an ideal photovoltaic material because the major defects such as ‘dangling bonds’ associated with a-Si which act as recombination centres are neutralized due to the presence of hydrogen. The hydrogen content is generally 5 to 20 atomic percent and by changing hydrogen content, the electronic or optical properties could be manipulated over a wide range.

Hydrogenated amorphous silicon with direct optical band gap of 1.7 eV is about one hundred times more effective in absorbing the Sun's radiation(optical absorption coefficient > 10^5 cm^{-1}),- a film of one micrometer thick would absorb as much as a 100 micrometer thick crystalline Si. This means that material usage is reduced and hence the cost. The first PV device based on a-Si has been fabricated by Carlson and Wronski in 1974 (Ref.26a). The production process steps for a-Si are the following:

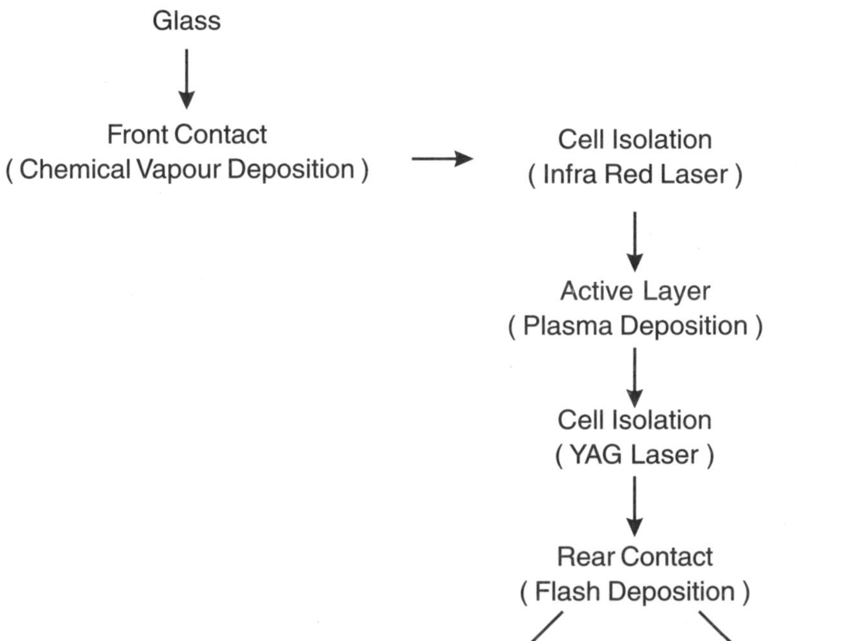

Doping severely degrades the lifetime of minority carriers in a-Si solar cells; hence p-i-n structure is used for the cell formation so that the solar light is absorbed in the undoped i-region and also generate electron-hole pairs in the same layer. A view of a typical p-i-n structure is shown in Fig.2.6a.

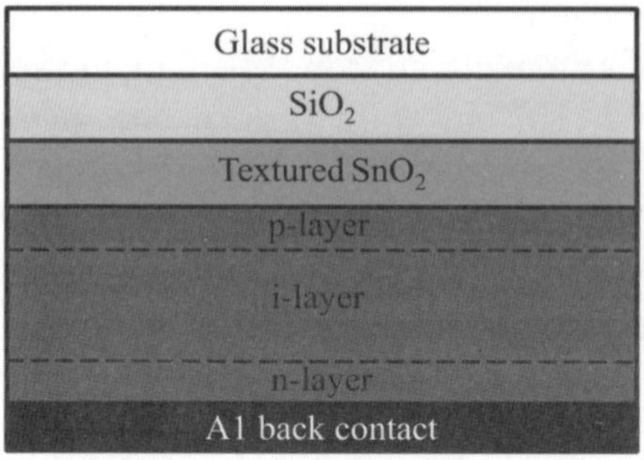

Fig. 2.6(a) *A single junction cell*

The modules made with p-i-n structure cells of efficiency > 12% when exposed to sunlight over a period of months, their efficiency degrades to less than 5% due to Staebler-Wronski effect (Ref.24). But it is possible to absorb the solar spectrum more efficiently and improve cell stability by using multiple p-i-n structures with different energy band gap i-layers to produce double or triple junction structures. Generally, silicon is alloyed with germanium to produce narrower or graded band gap layers. The schematic 'double-junction' and 'triple-junction' structures are shown in Figs. 2.6b and 2.6c.

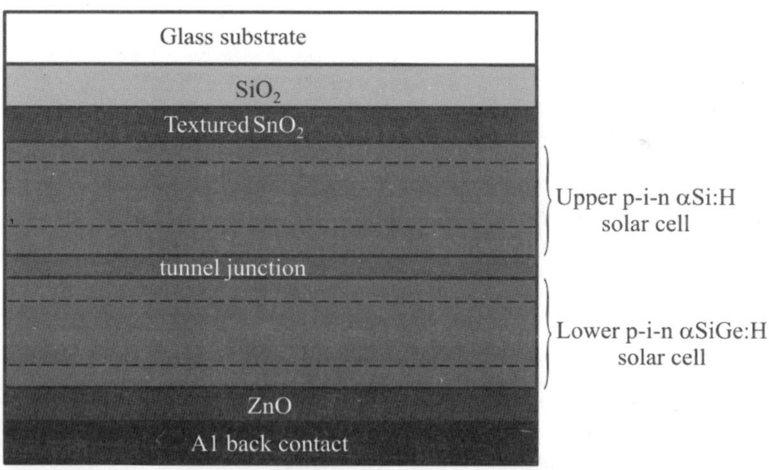

Fig. 2.6(b) *A double junction cell*

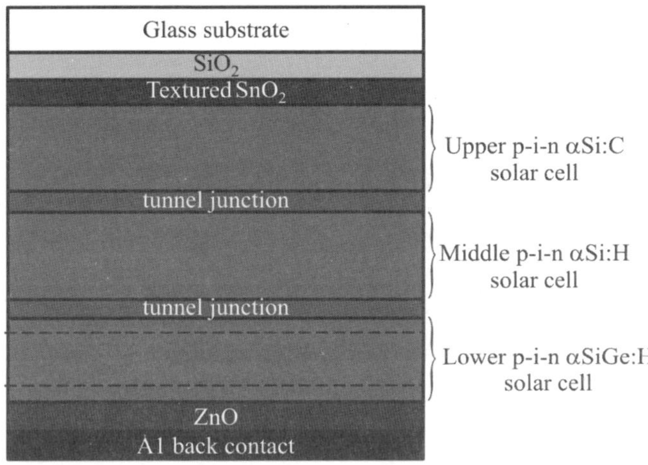

Fig. 2.6(c) *A Triple junction amorphous Si solar cell.*

At present the highest reported stabilized efficiency for a single-junction module is >8%, for a double-junction >9.5% and for a triple-junction module >10% (Ref. 24).

Guha prepared a triple-junction cell structure (Ref. 2.6b) where cells of different band gaps are stacked together. The top cell, which captures the blue photons, uses a-Si alloy with an optical gap of ~1.8eV for the intrinsic *i* layer. The *i* layer for the middle cell is an amorphous silicon-germanium (a-SiGe) alloy with about 10-15% Ge. The optical gap is ~1.6eV, which is ideally suited for absorbing the green photons.

The bottom cell captures the red and infrared photons and uses an *i* layer of a-SiGe alloy with about 40-50% Ge, corresponding to an optical gap of ~1.4eV. Light that is not absorbed in the cells gets reflected from the silver/zinc oxide (Ag/ZnO) back reflector, which is usually textured to facilitate light trapping. The cells are interconnected by the heavily doped layers that form tunnel junctions between adjacent cells. A roll-to-roll automated process has been developed for manufacturing solar cells on stainless steel. Rolls of stainless steel, a mile and a half (2500 m) long, 14 in (36 cm) wide, and 5 mil (125 μm) thick, more in a continuous manner in four machines to complete the solar cell fabrication. More work on a-Si modules are found elsewhere too. (Ref. 26c,d)

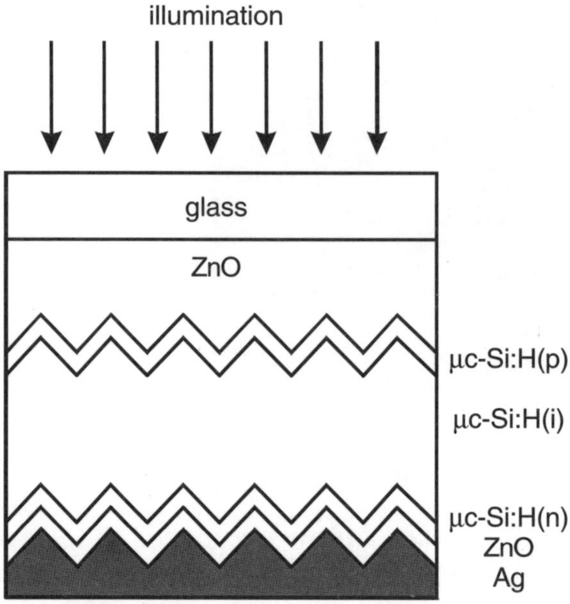

Fig. 2.6d *Schematic view of the layer structure of μc-Si:H thin film solar modules. The total thickness of all thin film (including TCO, Si, Ag) is < 3μm.*

Microcrystalline silicon:

If the gases used in PECVD for the deposition of a-Si are diluted in hydrogen, the deposited layer consists of regions of crystalline Si immersed in an amorphous matrix (Ref. 16). This two-phase material is called 'microcrystalline Si (mc-Si) or 'nanocrystalline Si'. The physical properties of this material especially with regard to stability under sunlight are similar to those of mono- or multi-crystalline Si, rather than amorphous Si. A schematic view of the mc-Si:H thin film module, shown in Fig. 2.6d has been developed.

The development of microcrystalline silicon (μc Si: H) thin film solar modules on textured ZnO coated glass with the following process steps is another step in progress.

1. TCO (ZnO) preparation.
2. TCO laser patterning.
3. deposition of the μc Si: H p-i-n structure by PECVD.

4. silicon laser patterning.
5. ZnO/Ag back contact sputtering.
6. back contact laser patterning.

Hybrid tandem solar cells combining stacks of low band gap micro-crystalline silicon and high band gap a-Si are developed. These are called micro-morph cells (Ref. 26e). Several groups have reported micromorph dual junction modules deposited by VHF (Refs. 26d, 31a, b). Results indicate a maximum efficiency of 12% which is very close to maximum stabilized active area efficiency of 13% obtained for triple a-Si/a-SiGe based cells (Ref. 32a).

These hybrid micro-morph devices and modules rival triple-junction a-Si in terms of stability and efficiency (Ref. 32b).

HIT (Heterojunction with Intrinsic Thin layer) Cells:

Another prominent high efficiency comericial cell is the HIT cell shown in Fig. 2.7a. Both crystalline and amorphous silicon are combined in this structure (Ref. 27, 28).

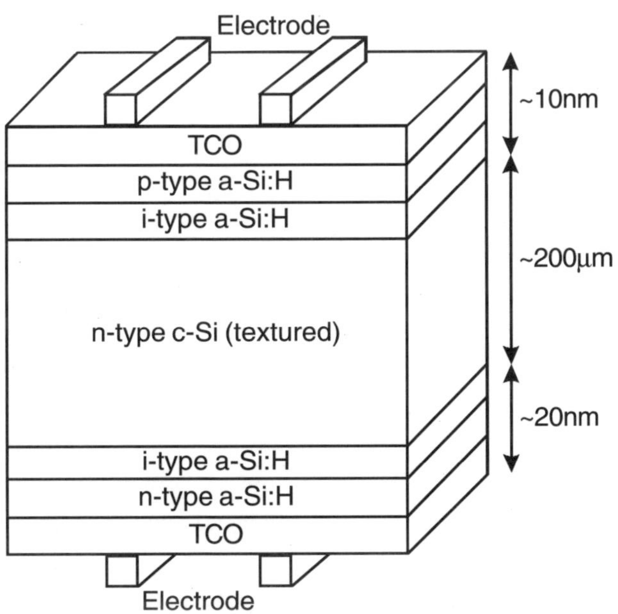

Fig. 2.7(a) *HITCell of Sanyo Electric*

The HIT solar cell structure is composed of a randomly textured solar-grade Cz silicon substrate sandwiched between p/i a-silicon layers on the front side(illuminated side), and i/n a-silicon layers on the back side as a back surface field(BSF) structure. Transparent conducting oxide (TCO) layers and metal electrodes are formed on both doped layers. The TCO layer on the front side whose thickness is optimised , serves as an anti-reflection layer too. The finger structure is used for the back side electrodes also, to reduce the thermal and mechanical stress in the cell. This makes the cell suitable for use as bifacial modules.

For example, Sanyo has recently announced 21% efficiency on the 10cm × 10cm HIT Cell and with efficiency upto 17.3% on 80cm × 120cm size module (Ref. 28). It also shows an excellent suppression of surface recombination,- the most important feature of high efficiency cells.

This structure has number of advantages: Potential for high efficiency, an excellent surface passivation of c-Si surface defects by high quality intrinsic a-Si layer, low temperature processing (except substrate production, all steps at less 200° C), better temperature co-efficient compare to c-Si cells resulting in a higher output power at high temperatures, reduced energy payback time and less cost compared to crystalline silicon cells (Ref. 10, 29). The symmetrical cell structure and the low temperature processes help to decrease the cell thickness.

Sanyo has developed a 200W HIT Double Bifacial module of size 1.66 m^2/ 24 kg, with an efficiency of 16.2%. It is structured by inserting a front and back symmetrical cell between two glass plates, enabling power to be generated from both sides; both direct sunlight and reflected sunlight incident on the module simultaneously generate electricity. This type of structure increases output power by 5% and by fixing them perpendicular to south, the annual output power increases by 34%(Ref. 28). A schematic diagram of a double sided module is shown in Fig. 2.8. The double glass structure enhances aesthetics and allows brilliant light and shadows to shine through the panels. The panels have a silver anodized aluminium, double-wall frame for extra strengths. They are safety rated for wind, hail and fire.

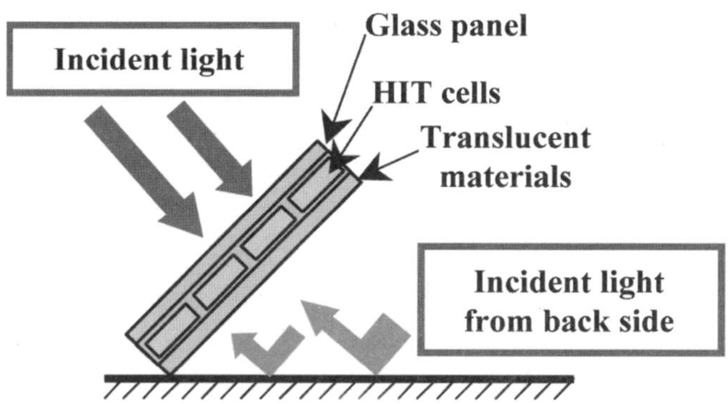

Fig. 2.8 *A schematic diagram of a double sided module*

These cells offer big cost benefits because of low material consumption and the advantage of continuous production on inexpensive substrates. The size of a single crystal silicon solar cell is limited by the diameter of the single crystal ingot, but thin film cells can be made in almost any size and shape by continuous process. The demerit which actually delayed its usage in power generating systems, is its degradation over a period of time on exposure to solar radiation. This has been mostly overcome by engineering the layer thickness and by using multiple junction structures.

2.5 TRANSFER TECHNIQUES FOR THIN FILM SILICON SOLAR CELLS

Several thin film silicon technologies (Refs.33a-c, 34) have been investigated with the aim of producing cost-effective and efficient solar cells using silicon substrates (Ref.35) as well as non-silicon substrates (Ref.36). The silicon substrates are either a low electronic quality coated with an epitaxial film of higher quality silicon or silicon sheets and ribbons which can be directly used as absorber layers. The non-silicon substrates are coated with a small amount of silicon before proceeding with the cell formation. There are different approaches in the usage of non-silicon substrates, to obtain silicon films of different nature and character. The recent one is layer transfer (or lift off) technique (Refs. 31 & 37).

This approach is expected to result in thin film solar cell efficiencies in the range of 18 to 20% and establish itself as an efficient and cheap Si-based thin film technology in the near future (Ref.38). The layer transfer techn iques are of two types: 'porous silicon (psi)' process (Ref.39,40) and 'epilift' process (Ref. 41). The psi-process utilizes a porous surface Si layer of 1 to 5 micrometer thick on the substrate formed by anodic etching of the wafer surface to permit detachment of the film. The porous film acts as a seed for the growth of an epitaxial p-n junction which is transferred to a low cost substrate. This process was first demonstrated using an ion-assisted deposition (IAD) on subs trates that are texured with regular inverted pyramids (Ref. 42). Sony corporation used a sacrificial porous silicon layer and prepared a 12 micrometer thick cell of 4 square cm area with an efficiency of 12.5% and V_{oc} of 623 mV (Ref.39b). This group, in order to develop mass production technology and to improve the conversion efficiency, has fabricated integrated-type 10 micrometer thick single crystal silicon solar cells with the light trapping structure (Ref.43) which needs further improvement.

This porous silicon route is slightly modified by forming a porous double layer of two levels of porosity (Ref.37,38). The silicon film with low porosity becomes a monocrystalline silicon seeding film, on thermal annealing, for high temperature epitaxial growth. The high porosity Si film becomes a sacrificial film and detaches the epitaxial layer from the starting substrate. The monocrystalline film includes voids formed during the process of crystallization and hence the actual nature is only quasi microcrystalline. The device processing of 'transfer' solar cells is described by Bergmann (Ref.38 a,b) and cells of efficiency,15.3% have been produced. This technique would heavily save on materials because the silicon consumption is only one tenth to one-hundredth of what is consumed in wafer technology. The issues related to throughput, yield and scaleability, however, are to be investigated for the lift off process which has demonstrated a great potential for producing cheap and efficient Si solar cells.

2.6 THIN FILM SOLAR CELL TECHNOLOGIES

With the emergence of thin film second-generation PV cell technology which has the triple advantages of minimal material use, large-scale manufacturing facility and low energy consumption during production, thin film solar cells based on copper indium selenide (CIS) and its modifications, and cadmium telluride (CdTe) have come to the stage of commercial production though certain critical issues related large-area module production are under intensive study.

$CuInS_2$ is another attractive material and significant work has been done mostly by European and Japanese groups on $CuInS_2$ based thin film solar cells.

2.6.1 CIS-based Solar Cells

Copper indium diselenide and its alloys have been the subject of intensive study in the last three decades. This material has direct energy band gap and high optical absorption which helps to minimize layer thickness and to tailor the desired optical response and the electronic behaviour. Both n- and p- type materials are possible and the cells have stable electro-optical properties. CIS cells are hetero-junctions between cadmium sulphide and CIS. In 1974, Wagner et al produced a $CuInSe_2$/CdS cell with an efficiency of 12%(Ref. 44). The composition, thickness and interfaces of each cell layer are engineered for the highest performance, stability and reliability. It was realised that incorporating gallium into CIS to produce CIGS resulted in the widening of energy band gap to 1.3eV and an improvement in material quality.

The performance of CIGS cells depends on Cu/(In+Ga) ratio and Ga/In profile of the CIGS films. The efficiency of these cells are improved by band gap grading in the CIGS film (Ref.45). The band gap of the CIGS films can be altered by controlling In and Ga; hence, the band gap grading along the depth of the film can be achieved by monitoring the composition profile of Ga and In.

The pioneering investigations of Kazmerski (Ref.46) showed a cell efficiency of 6% as early as 1977. The potential for low-cost manufacturing of these cells is very high as indicated by a variety of preparation techniques used to fabricate cells/modules. Out of these techniques, three methods have dominated for large scale production: (i) co-evaporation of elements (Ref. 47-50), (ii) selenisation of elemental precurser layers (Ref.51,52), and (iii) stacked elemental layer (SEL) processing (Ref.53).

The typical structure of CdS/CIGS thin film solar cell used in all methods is shown in Fig.2.9. The first method yields devices with highest performance while the second one is preferred for large-scale production.

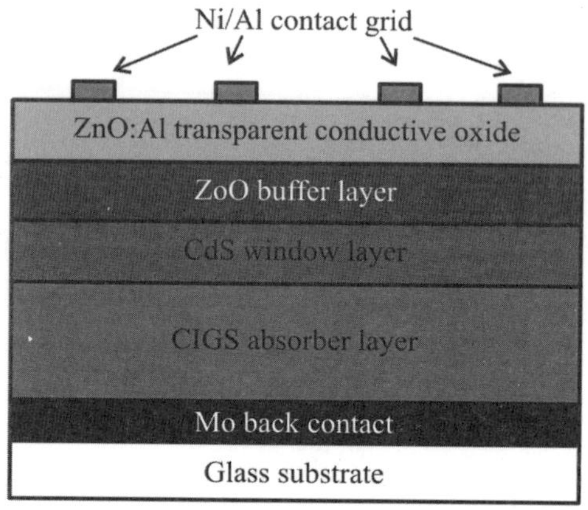

Fig. 2.9 *Cross sectional view of CdS/CIGS thin-film solar cell.*

Both the methods have to be processed at temperatures >500°C to enhance crystallization and grain growth. The passivation of grain boundaries and grain growth are attributed to the presence of sodium (Refs.54-57), which diffuses into the absorber layer either from the glass substrate or introduced chemically by evaporation of a Na-compound. This will give rise to increase in efficiency by 1 to 2%.

CdS is found to be the best window material. But alternatives, ZnS, ZnSe, In_2S_3, (Zn, In)Se, Zn(O,S) and MgZnO (Refs. 58, 59) can also be used.

The main producers of CIGS modules are Wurth Solar, Avanics and Global Solar (Ref.10), though there are other players in the field.

Significant progress has been achieved in terms of reliability and efficiency;- 19.2% efficient research cells and 21.2% on concentrator cells (14x) fabricated with three-stage co-evaporation process are reported (Ref. 47, 48). The module efficiencies established are 16.6% on an area of 16 square cm (Ref. 60) and 12.7% on an area of 736 sq.-cm (Ref. 49).

Uptodate reviews in the development of thin film devices are given by Ullal and co-workers (Refs. 61, 62) and others (Ref. 63, 64). CIGS cells with efficiencies of 19.5% (Ref.65) and modules with efficiencies of 13.4% (Ref.66) have also been reported.

The process development of CIGS modules for mass production developed by ZSW, Stuttgart and WURTH Solar (Refs. 67, 68, 50) has given excellent results. Wurth solar is the first CIS company successfully established in the PV market. The main objectives of this process are to enhance the reliability and to reduce the costs of production. The base-line processes to fabricate modules of 30 cm × 30 cm are given in Table. 2.1. The process step, technology used and the parameters controlled in the process are shown in the table. All vacuum deposition processes are performed in one-step mode. All process steps are realized with a cycle time of 15 minutes or less. About 200 modules are produced in five batches continuously; and each batch is performed in a nonstop 3-shift processing.

Table 2.1 *Base-line processes at ZSW for CIGS modules*

Process Step (film thickness)	Process Technology	Process Control
1. Soda-lime glass (2-4 mm)/ Mo (0.5-1 micrometer)	DC magnetron sputtering	thickness, sheet resistance, adhesion
2. Mo patterning	Nd-YAG laser patterning of stripes	electrical insulation
3. $Cu(In,Ga)Se_2$ (2 micrometer)	one step co-evaporation in-line	atomic absorption spectroscopy, x-ray fluorescence spectroscopy
4. CdS (50 nanometer)	chemical bath deposition	visual inspection
5. i-ZnO(50 nanometer) sputtering		RF magnetron thickness, sheet resistance, transmittance
6. CIGS/CdS/i-ZnO	mechanical patterning	visual inspection scribing
7. ZnO:Al (1 μm)	DC magnetron sputtering	thickness, sheet resistance, transmittance
8. ZnO patterning	mechanical scribing	I-V curve, dark and illuminated
9. Contacts	bus bars	visual inspection, resistance
10. Module packaging	EVA encapsulation	visual inspection, adhesion

The best modules produced have demonstrated efficiencies close to 13% with an average of 11.3%.The CIGS deposition is the most critical aspect and demands accurate control of evaporation rates of Cu, In, Ga and Se. Hence the evaporation rates of Copper, Indium and Gallium sources are controlled by atomic absorption spectroscopy and the Selenium source is temperature controlled. The composition of these co-evaporated films are found to be very reproducible with a deviation of 1%. The productivity of the process and the reliability of the product are examined for commercialization.

The yield and failure-analysis performed on one batch of 108 modules produced in one sequence have shown yields of about 90%. Processes which are compatible to metal foils, thin glass and polymer foils as substrates are being developed because of then flexibility and lightweight qualities. Process steps for a commercial production line to produce modules of size, 120 cm × 100 cm are established and are under optimization for establishing a pilot line of capacity, 1 MW/year. This group presented (Ref. 69a), the current status of 60 cm × 120 cm CIGS module production in the Wurth Solar Pilot Line and new developments at the ZSW, such as product-friendly 30 cm × 30 cm flexible modules and 120 cm long evaporation sources for faster module production. Dimmler has reported (Ref. 69b) commercial production of 15 MW/year in 2007, which has been scaled to 30 MW/year from July 2008. Improvements in overall process yield by continous process optimisation and process control, and reduction of cycle time and other innovations are envisaged to take the average module efficiency to >14%.

The Japanese group (Ref. 70) has developed another production process in which an in-line evaporation system is equipped with a temperature detector to monitor the composition of gallium and indium. The Ga to In ratio influences the band gap of the material and hence the efficiency of the cell. The cell structure is ITO/ ZnO/ CdS/ CIGS/ Mo/ Glass. The CIGS films are deposited by sequential three-stage process,- the Mo, ZnO and ITO films are deposited by RF sputtering and CdS films by chemical bath deposition.

In the first stage, In-Ga-Se precursors are deposited on a moving Mo coated glass substrate at a temperature of about 350°C. In the second stage, Cu and Se fluxes are exposed to the In-Ga-Se precursors at a substrate temperature of about 500°C to form Cu-rich CIGS films. In the final stage, small amount of In, Ga and Se fluxes is added to the Cu-rich CIGS films maintaining the same temperature as in the second stage to obtain slightly Cu-poor CIGS films. The cooling of the substrate is measured using a radiation thermometer through the viewing port located away from the heater. The depth profiles of Cu, In, Ga,

Se and Mo are determined using Secondary ion mass spectroscopy and the composition of the films by Energy Dispersive X-ray spectroscopy.

The schematic arrangement of the in-line evaporation system is shown in Fig. 2.10a. The CIGS sub modules are fabricated on 10 × 10 cm glass substrates. Mo layer (back contact) is deposited by DC-sputtering; the CIGS films are deposited as described above; CdS film as buffer layer is formed by chemical bath deposition; and the window layer, ZnO film, is deposited using RF-sputtering. ITO film which acts as transparent conducting electrode is prepared by RF-sputtering. The ZnO, CdS and CIGS films are patterned by mechanical scribing for the interconnection. The current-voltage measurements have shown an efficiency of 12.6% and fill factor of 69.9% with an aperture of 81.5 sq.cms. The optimization of band-gap grading may further enhance the module performance.

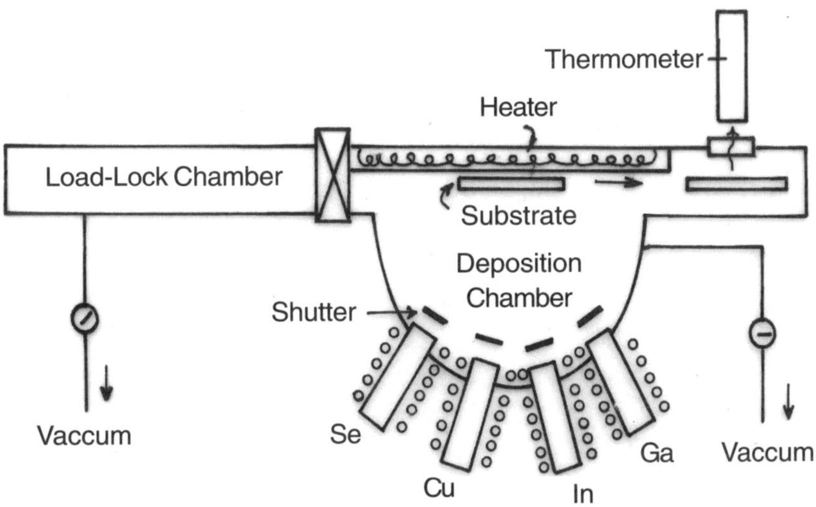

***Fig. 2.10a** Schematic of in-line evaporation system*

Siemens Solar which has introduced CIS modules into the market in 1998, has further improved the formation of absorber layer resulting in the improvement of module efficiency (Ref. 71). Highly efficient mini-modules 5x10 cm based on Cu(In, Ga) $(Se, S)_2$ chalcopyrite absorbers are fabricated with good reproducibility.

The solar cell grown on 10 cm × 10 cm glass substrates has a structure, Glass/Mo/Cu(In,Ga)(Se,S)$_2$/CdS/ZnO:B or ZnO:Al. The absorber layer is grown in a two step process: a Cu-Ga-In-Se precursor deposition, followed by rapid thermal processing (RTP) in sulphur containing atmosphere. The open-circuit voltage of the photo-generated charge carriers increase with sulphur concentration in the absorber film. An efficiency of 13.3 ± 1.5% has been achieved on laminated mini-modules before lamination. The improved efficiencies are attributed to the following factors:

(i) *RTP-absorber film formation in sulphur atmosphere,*
(ii) *tight process control of Na-impurity concentration,*
(iii) *Cu to (In+Ga) precursor ratio and*
(iv) *CdS-layer thickness.*

The potential for much higher efficiency is demonstrated by a champion mini-module with 14.7%; though the targeted efficiency on mass production is higher than 15%.

Swiss group (Ref. 72) has developed two 'lift-off' processes to fabricate very thin CIGS solar cells on flexible polymer films,- the flexible and lightweight modules are more attractive both for terrestrial and space applications. In the first 'lift off' process, the solar cell is prepared on polyamide coated glass substrates.

A buffer layer of NaCl is evaporated on the substrate; then a special type of polyamide film is spin-coated and cured at a temperature of 400-450°C. Cells of 3cm × 3cm and 5cm × 5cm are developed with an efficiency of 12.8% under AM 1.5 illumination conditions. After the entire cell structure is grown, the buffer layer is dissolved in water and the cell is lifted off from the substrate.

The growth temperatures are restricted to less than 500°C; however the potential to increase the area of the cell is high. The second process is designed to grow cells at temperatures greater than 550°C; the cell structure is processed directly on the substrates coated with a buffer layer of NaCl or NaF or BaF$_2$. On the top of the cell structure, a transparent polymer layer is coated. The glass substrate is removed after entire processing by dissolving buffer layer in water. This flexible solar cell has a structure of polymer/ ZnO:Al/ ZnO/ CdS/ CIGS/ Mo. Mo layer is grown by dc sputtering, the CIGS layers are deposited by co-evaporation of Cu, In, Ga and Se at a substrate temperature of 400-450° C, and the CdS layer by chemical bath deposition.

The conducting front layer of ZnO/ZnO:Al is deposited by RF sputtering at room temperature. NaCl is chosen for the buffer layer because it can provide Na to CIGS layer. The demonstrated efficiency is 8.2%. The two structures are shown in Fig. 2.10b and Fig. 2.10c.

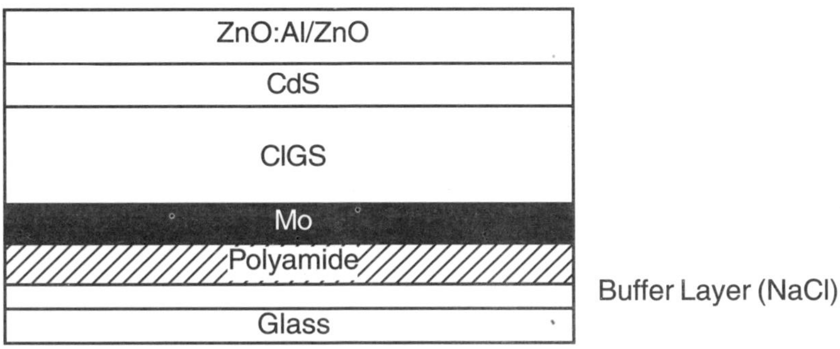

Fig. 2.10b *CdS/CIGS solar cell structure on a POLYAMIDE coated glass substrate with NaCl buffer layer*

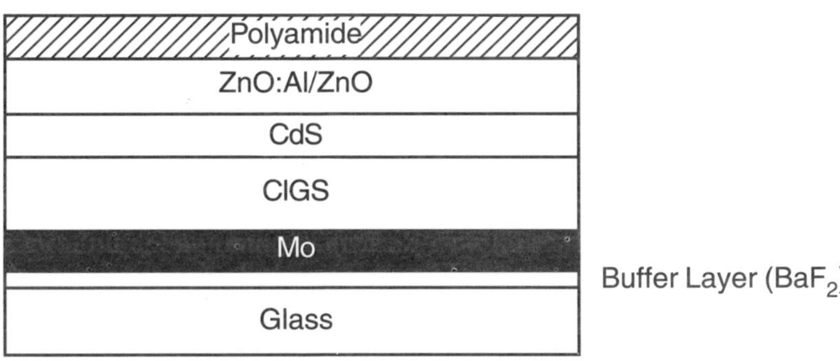

Fig. 2.10c *CdS/CIGS solar cell structure on glass with* BaF_2 *buffer layer*

This novel lift-off technology has immense potential to develop 'very thin' solar cells utilizing thin film technologies and to achieve efficiency higher than 15% on large area cells/modules. This process can be adopted for in-line or roll-to-roll deposition methods for industrial manufacturing of large area modules with high throughput and reliability.

2.6.2 $CuInS_2$ based Solar Cells

$CuInS_2$ is a good absorber material for thin film cells. Several papers have been published on these cells since early '90s (Ref. 73, 74). The advantages of this technology are (a) Se-free, (b) good homogeneity, (c) very fast absorber formation, (d) flexibility in the preparation methods and preparation parameters, (e) Ga and Ag incorporation making the device suitable for a particular application, (f) high open circuit voltage and low current density, and (g) feasibility of monolithic integration. However, further optimization is necessary on aspects, such as 'increased efficiency', Cd-free devices, and 'alternatives for the etching of the absorber'. The other issues which need to be addressed are long-term stability, scaling-up the production and strategy for commercial exploitation.

At present, cells prepared by different approaches using industrially relevant processes have shown efficiencies around 12%. It is observed that the sequential evaporation processes using sulfur vapour result in slightly better devices than those prepared by co-evaporation or H_2S sulfurisation.

At Northumbria, Zn S_{1-x} Se_x layers are developed for use as buffer layers to Cu In S_2, because this system creates the possibility of lattice matching across the junction (Ref. 75, 76).

Tremondous research activity at Hahn-Meitner-Institute, Berlin and other Centres on CuInS2 - based solar cells has been reported. From a host of publications available in the literature, a few selective recent papers, by R.Scheer et al (Ref. 76b), HJ Lewerenz (Ref. 76c), R.Klenk et al (Ref. 76d), Jo Klaer et al (Ref. 76e), B.Berenguir et al (Ref. 76f), T.Theresa John(Ref. 76g), B.Asenjo (Refs. 76h, i), S.Nakamura et al (Ref. 76j) need mentioning.

2.6.3 CdTe- based Solar Cells

Cadmium telluride is an excellent solar cell material with highest theoretical conversion efficiency. With a direct energy bind gap 1.5eV and high optical absoption coefficient, CdTe is a promissing PV material (Refs. 77, 78) and a few thick microns are enough to absorb most of the incident light. Several preparation techniques such as vacuum evaporation, spraying, screen-printing and electro-deposition have been initially used to produce good solar cells. Several cell structures have been investigated to improve cell stability and efficiency (Ref. 79-81).The conventional cell structure for many years has been SnO_2/CdS/CdTe shown in Fig. 2.11.

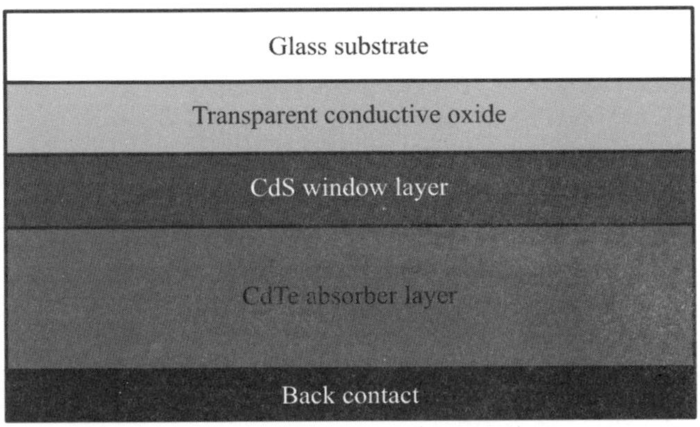

Fig. 2.11 *CdS/CdTe thin-film solar cell.*

In the structure shown, the layers were deposited onto the glass substrate coated with a transparent conductive oxide (TCO), with the sequence: CdS window layer, the CdTe absorber layer and the back contact. This structure is also referred to as 'superstrate configuration device' (Ref. 29). Several methods have been used to deposit CdS layer, but for commercial cells either close-spaced sublimation (CSS) (Ref.82) or chemical bath deposition (Ref.83) have been used. The CdTe absorber layer, 3-10μm thick, can be deposited using CSS (Refs.84a,b,85) or physical vapour deposition (Ref.86) or electrodeposition (Ref. 87) or spray pyrolysis (Ref.88a). More details on CdTe processing and devices are given elsewhere (Refs. 64, 88b). The CSS is a thermal evaporation method, in which the 'source-substrate' distance is typically a few millimeters; an inert gas atmosphere is maintained between the source and substrate. Due to the high deposition temperatures, the CdTe layer contains dense, facated grains. To produce high efficiency device, the deposited layer is activated in the presence of $CdCl_2$. This treatment is known to recrystallise CdTe layer (Ref.89), passivate grain boundaries, and promote interdiffusion of the CdS and CdTe at interface. To deposit back contact to CdTe-layer is difficult because the work function of CdTe is higher than all metals. So, the surface is etched in a solution of bromine in methanol or HNO_3/HPO_4 acid solution and depositing Cu-Au alloy or Zn-Te:Cu. This process creates a thin p+ layer that charge carriers can tunnel through. The presence of copper causes degradation of the cell because it can easily diffuse into CdTe-layer. Alternately, a very low band gap material such as Sb_2Te_3 followed by molybdenum or tungsten (Ref.90) can be used which does not require surface etching.

The cells are relatively stable although humidity affects to some extent. The spraying and screen-printing are the techniques with high economic potential. The toxicity of cadmium raise two main problems,– possibility of production hazards and environmental pollution. In spite of these demerits, there has been significant progress in developing low-cost manufacturing processes for rapid commercialization of CdTe cells.

Solar cells of 16.5% efficiency have been prepared by Close-Spaced Sublimation (CSS) method (Ref. 91, 92); Commercial scale modules of 8 to 9% efficiency have been reported by a few manufacturers (Refs. 83, 85).

At NREL, USA, a manufacturing-friendly process with a device structure of Cd_2SnO_4/Zn_2SnO_4/CdS/CdTe has been developed (Ref. 83). Several production disadvantages inherent in the manufacturing process of the conventional SnO_2/CdS/CdTe structure such as deposition of SnO_2 and CdTe layers at high temperatures using CVD and CSS processes, and the intermediate CdS layer at a lower temperature using chemical bath process are addressed in this NREL's process. The heating and cool-down aspects are time-consuming and expensive. Due to the changes effected in the new method, the performance and reproducibility are significantly improved. For instance, the Cd_2SnO_4 (or CTO) layer has higher conductivity and optical transmission compared to SnO_2 layer; the Zn_2SnO_4 (or ZTO) layer acts as a buffer layer to maintain high V_{oc} and FF; and the CdS layer thickness is reduced to improve I_{sc}.

The manufacturing process is simplified by depositing CTO and ZTO layers and CdS layers in pure argon at room temperature by the highly matured RF magnetron sputtering using hot-pressed oxide targets. The CdTe layer is deposited by CSS method at 570°C-625°C, the only heating part in the entire process. The recrystallizations of the CTO, ZnTO and CdS layers and the interdiffusion at the three interfaces, CTO/ZTO, ZTO/CdS and CdS/CdTe interfaces, are completed during CdTe deposition. These samples are then treated in $CdCl_2$ vapour at 400-420°C for 15 minutes.

The cells fabricated by this process have shown an efficiency upto 16.5%(Ref. 84a) with a FF of 74.95%. This is the highest efficiency ever reported for thin film CdTe cell wherein TCO and CdS are deposited at room temperature. It is entirely a 'dry process' with no liquids involved. The process time is relatively reduced while increasing' throughput'. The yields' are also high because only two compatible deposition techniques are involved. The CdTe cells produced with this process have good uniformity, reproducibility and device stability.

Another low-cost in-line fabrication system has been developed by German group (Ref. 93) by looking closely at the complex character of the materials involved in the fabrication of CdTe cells. It is a combined 'in-line sputtering and CSS equipment' consisting of four independent processing chambers. The substrates are 10 cm × 10 cm and the layer structure of the cell is Ni/ ZnTe/p-CdTe/n-CdS/ITO+TO/ glass. The system design is such that it avoids interruption of vacuum and long cooling cycles during the process.

The transparent contact layer (TO on ITO) is prepared by RF sputtering, and the p-CdTe and n-CdS layers are prepared by CSS technique under a defined inert or reactive gas (Ar, N_2,O_2) atmosphere at working pressures of 100 m bars. While oxygen enhances the p-type character of CdTe, the inert gas atmosphere influences the grain size of CdS and CdTe. The p-ZnTe layer made by CSS acts as a diffusion barrier for the metal back-contact, Ni, and creates no barrier on p-CdTe layer. The metal contact is prepared by DC sputtering. The CdS/CdTe layer system is activated by annealing in the presence of $CdCl_2$ or hydrochlorine or chlorine gas, which initiates the recrystallizations of the chalcogenide by removing structural defects and forming an intermixing layer of $CdS_xTe_{(1-x)}$ at the CdS/CdTe interface. The thickness of the intermixing zone is limited to a few nano-layers only which helps to decrease the lattice mismatch and the interface defects. The post-growth activation has significant effect on the V_{oc}, I_{sc}, FF and conversion efficiency. The schematic diagram of the in-line system is given in Fig. 2.12.

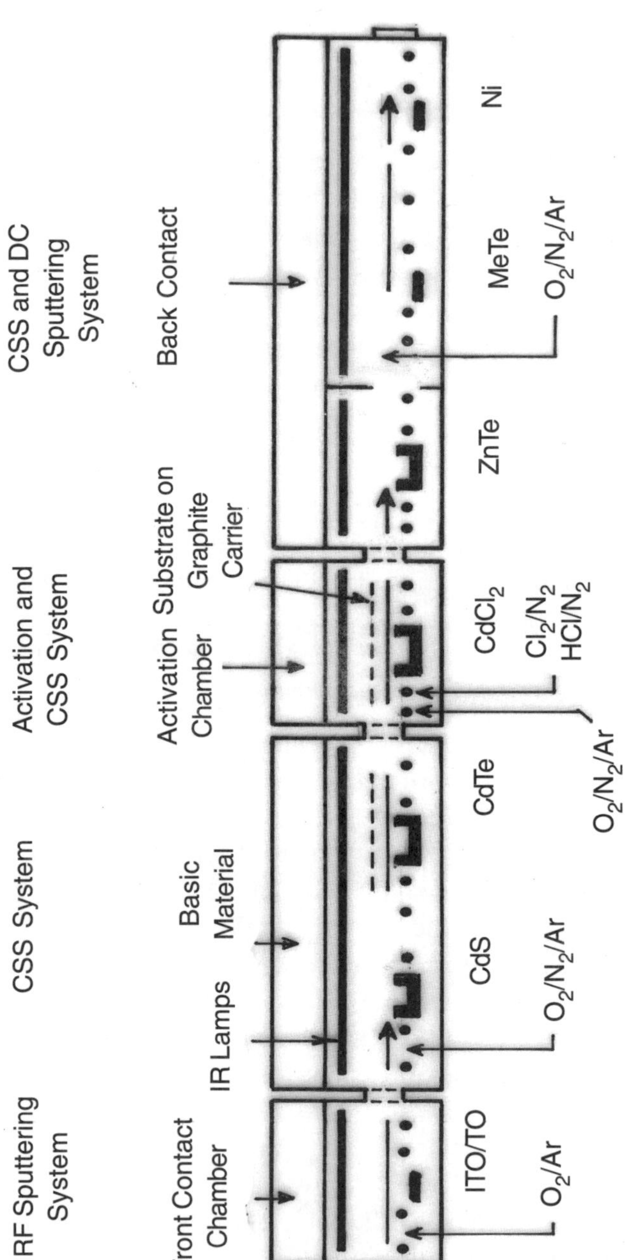

***Fig. 2.12** Schematic of in-line sputtering and CSS equipment for fabrication of CdS/CdTe thin film solar cell*

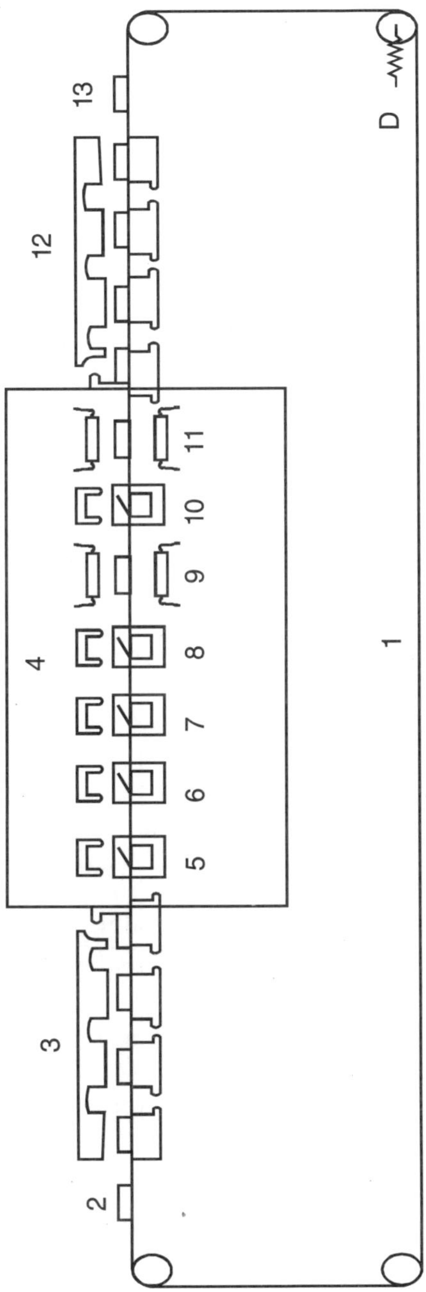

(1) Belt Conveyor (2) Substrate (3) AVA Seal (4) Vaccum Chamber
(5) Heating the Module (6) CdS deposition (7) CdTe deposition
(8) $CdCl_2$ deposition and heat treatment (9) $CdCl_2$ annealing and stripping
(10) Ohmic Contact Formation (11) Contact Annealing
(12) AVA Seal (13) Fabricated Cell

***Fig. 2.13** Schematic of Pilot System for CdTe/CdS Cells*

Another continuous in-line process for fabricating large volume CdTe solar cells has been reported (Ref. 94) recently. The initial cells from this process have shown efficiencies, 10.5 to 12.5% which have been repeatedly demonstrated. The NREL verified efficiency is 12.4% with a fill factor of 71.03% measured at 25°C within a variation of 1°C. The pilot-line is a continuous ,in-line system performing all fabrication steps in one vacuum zone. The structure is Glass/SnO_x:F/CdS/CdTe/carbon/nickel. (Fig. 2.13).

The different processes, namely, module heating, CdS and CdTe deposition, $CdCl_2$ deposition and heat treatment, annealing and stripping, ohmic contact formation and annealing are carried out in one sequence under one vacuum boundary. Metallization is performed outside of vacuum by spraying. The ohmic contact is prepared through the vapour deposition of a copper compound followed by annealing.

There is no antireflection coating. The substrate glass is unmodified soda lime glass. Outdoor testing of these modules have shown that copper-based ohmic contacts can achieve good stability. Cells of 3.6 in × 3.1 in size are produced,- each cycle in a processing time of two minutes. The system size is being scaled up to take 12 in × 12 in module. This continuous process is scalable, uniform and reproducible. It is merited by increased cell reliability, low production time and cost viability.
The leading manufacturer of commercial CdTe-based devices are Fist Solar,USA (Ref.95). The current production costs of First Solar are US$ 1.25/ Wp for a production line of 99MWp per year; the projected output for 2008 is 275 MWp per year with estimated cost below US$ 1/Wp barrier (Ref.96a). The company has announced in February 2009, that the manufacturing cost is reduced to $0.98/watt, breaking $1.0/watt price barrier (Ref.96b). The largest thin film installation using CdTe modules of First Solar was on the roof of a logistic center in Ramstein, Germany in 2007. This system is capable of generating 2.5 MWp of power (Ref.10).

A number of groups worldwide are actively involved in developing polycrystalline thin-film PV technologies. Table 2.2 summarizes the champion module aperture-area efficiencies made by several organizations worldwide (Ref. 97). A thin-film CIGS-based power module efficiency of 13% was achieved by several groups and thin-film CdTe power module efficiencies of 10%-11% were achieved by three groups. Several companies are scaling up manufacturing facilities worldwide to meet the growing market demands for low-cost reliable thin-film PV products.

Table 2.2: *Polycrystalline Thin Film Photovoltaic Modules (Ref. 97)*

Organization	Material	Area (cm^2)	Eff. (%)	Power (W)	Date
BP Solar	CdTe	8390	11.0*	92.5*	09/01
Wurth Solar	CIGS	6507	12.2	79.2	05/02
First Solar	CdTe	6612	10.1*	67.1*	12/01
Shell Solar-GmbH	CIGSS	4938	13.1	64.8	05/03
Matsushita	CdTe	5413	11.0	59.0	05/00
Global Solar	CIGS	7714	7.3*	56.8*	03/02
Antec Solar	CdTe	6633	7.0	46.7	11/01
Shell Solar	CIGSS	3626	12.8*	46.5	03/03
Showa Shell	CIGS	3600	12.8	44.15	05/03

* NREL confirmed

2.6.4 Dye-sensitized Cells

The dye-sensitized cells are considered as thin film cells. The principle of working is based on photosensitization of wide-band-gap semiconductors.

A wide-gap semiconductor with a large surface area is covered with dye molecules. When the light is incident, the light is absorbed in the dye molecules which are excited and the electrons from their excited state are directly injected into the semiconductor, without the need for transport of photo generated carriers within the dye. The ground state of the molecules has to be filled again so that the process goes on. This principle was reported by Tribtsch and Gerischer (Ref.98) and the device-structures by O'Regan and Graetzel (Ref.99,100). These are referred to as Graetzel cells and schematic structure is shown in figure 2.13a.

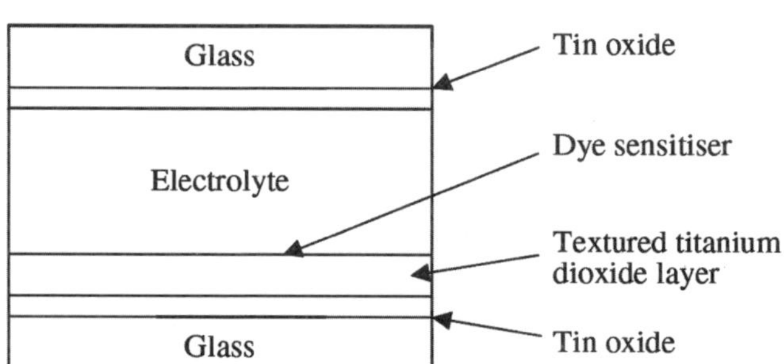

Fig. 2.13a *Graetzel solar cell*

The separation of optical absorption and charge separation are achieved in this device by the presence of a dye sensitizer as a light absorbing material with a wide gap semiconductor of nano-crystalline or mesomorphic nature (Ref. 101, 102). The dye sensitized cells are particularly well suited for the fabrication of tandem structures for an optimal utilization of solar specturm (Ref 103-106). In Germany, the company, INAP GmbH has been working on the development of dye-sensitized TiO_2 cells, and efficiencies of 7% on 30 cm × 30 cm areas have been reached. There has been a lot of research activity and new photosensitizing chemicals are developed (Ref. 107). An efficiency of 11.2% has been reported for a 0.25 sq.cm area device (Ref. 108, 109). The coversion efficiency over 15% has been obtained with tandem cell on a laboratory scale, and the scope to improve the efficiencies is great (Ref. 110). This is considered as a potential low-cost PV technology, not requiring very expensive high temperature and high vacuum processes. The fabrication is not energy intensive and can be produced on flexible substrates in a variety of appearances. They have an edge over silicon technology in providing power both indoor and outdoor due to their excellent performance in diffuse light.

An extensive review on the development and status of these cells is given by Gratzel (Ref. 110).

Organic solar cells:

These cells are made of thin layers of organic materials. The advantages of these cells over the conventional cells are easy preparation, low cost materials, low cost processing and low process temperature. But the efficiencies are comparatively low.

Another advantage is that they can be prepared on flexible plastic sheets/ foils, enabling them highly suitable for many applications and for mobile systems.

There is a distinct difference in the operating principles of conventional (inorganic) cells and the organic cells. In the former, the free charge carriers are directly generated while in the latter, the photon absorption is followed by the creation of excitons with a binding energy of around 0.3 to 1.0 eV. The required electric field to overcome this binding energy is not available in organic cells; hence, the excitons are separated by the interface between two different organic layers constituting an hetero-junction. The relative energies of these two materials need to be optimized to keep the excitons separated and to avoid the energy loss in the process.

The cell structure for most of these cells comprises the following: a glass substrate coated with indium tin oxide, photovoltaic organic material layer, and a metallic layer. The substrate acts as the transparent top electrode and the metallic layer acts as the back contact for the cell. With one absorber organic layer during the initial studies, the carrier collection and hence the efficiency were found to be poor. Later structures have used two organic layers.

There are two distinct types of organic solar cells: (i) Small molecule solar cells, which are processed in vacuum by physical vapour deposition (PVD), and (ii) Polymer solar cells, which are prepared by spin coating or ink-jet printing.

The absorber materials are broadly categorized as follows:

Phthalocyanines: ZnPc or CuPc are the commonly used materials in small molecule cells. They have high absorption coefficients in red region of the solar spectrum and are used as 'donors' in a hetero-junction. Since they have low exciton diffusion lengths, they are used in bulk structures.

Fullerene (C60): It is used as an 'acceptor' material. Due to their high diffusion lengths, layer of several tens of nanometers thick can be used in the PV cell. It shows strong absorption in the UV- and blue regions of the solar spectrum.

Oligothiophene: The IAPP, University of Dresden has recently studied these materials as absorber material in organic cells. They consist of thiophene rings in a chain and strong electron-withdrawing cyano-end groups. These molecules act as both donors and acceptors in organic solar cells, depending on the number of thiophene rings and the material combination in the device (Ref.111)

The hetero-junction between two highly absorbing materials has been the basis for the functioning of the organic cells. This type of junction is needed to separate the excitons which are strongly bound in organic semiconductors. Due to short diffusion lengths, the photoactive region of such cells is only a narrow layer on either side of the hetero-junction. The cells have to be much thicker than the active region to avoid recombination at the metallic contacts. This means, that there are high absorbing regions which do not contribute to the photocurrent.

These regions are replaced by transparent materials (wide-gap transport materials) at IAPP, University of Dresden, in their research activity leading to a cell structure: first electrode, transparent layer, photolayer(s), second transparent layer and second electrode as shown in the Fig.2.14(a).

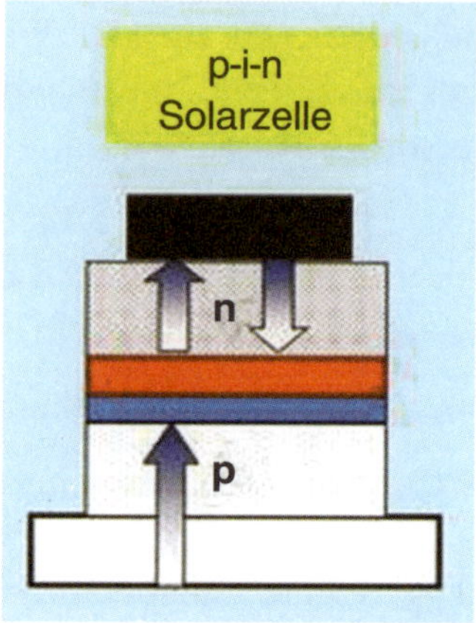

Fig. 2.14a *p-i-n hetero-structure for an oganic cell (Ref. 112)*

The advantage of this (p-i-n) design is that the solar cells only absorb the light in the photoactive region and also achieve very high internal efficiencies (which is crucial for developing tandem cells). The other important technological feature of this p-i-n concept is the 'controlled doping' of the wide-band gap transport layers. While the doping process is critical to inorganic cells, the organic solar cells are usually prepared without controlled doping, because of the difficulty to produce stable and reproducible doped organic layers. But, stable and reproducible doping in organic systems improves the device properties (Ref.112). For eg. using weakly acceptor-type matrix materials such as C60, conductivity values of higher than 10^{-5} S/cm are reached, high enough to make ohmic losses in n-doped transport layers of organic cells negligibly small.

Fig 2.14(b) shows the energy diagram of an ideal p-i-n hetero-structure for an organic cell. In this cell, only the intrinsic layer absorbs visible light. This layer can be a bilayer of two highly absorbing materials or a bulk-hetero junction (blended layer). The p- and n- layers are realized by doped wide-gap materials. The role of interfaces at the photoactive layers is very crucial because the transition of photo generated charge carriers from the photoactive region to the respective transport layer must not only be barrier-free but should occur without free energy loss. This is possible with a good energy level-alignment.

In addition, the injection of photo generated carriers into the transport layers as minority carriers have to be prevented by high barriers.

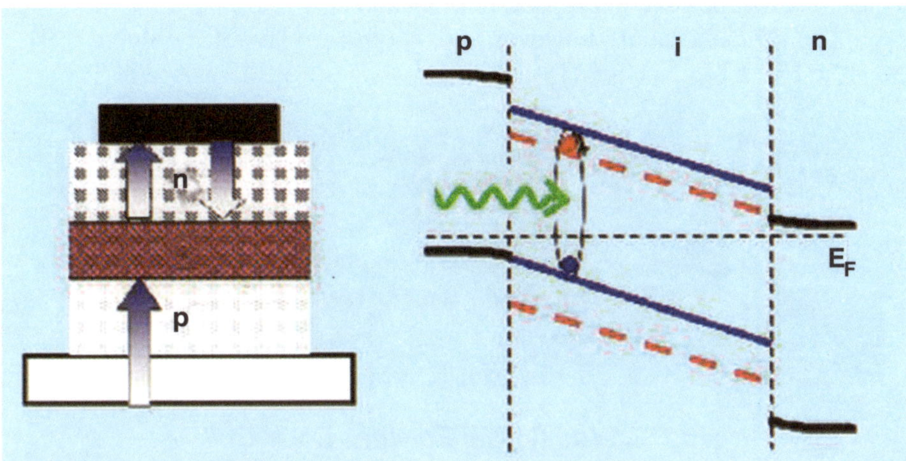

Fig. 2.14b *Energy diagram (Ref. 112)*

A major advantage of wide-gap materials lies in optimizing the cells in terms of thin film optics. The thickness of the photoactive layer can be varied to achieve high internal efficiency. Thickness variation of the wide gap layers also allow placing of the photoactive layer at the maximum of the optical field distribution inside the device leading to an increased absorption.

This concept can be extended to prepare stack of cells with integrated light trapping structure that improves the absorption while achieving high power efficiency of 8 to 10%.

Recently an innovative design responsive to near infra red region of the spectrum has been reported (Ref 113). The device uses C60, Copper and Tin phthalocyamines (CuPc and SnPc). This design has demonstrated that significant power can be obtained from the IR and near IR regions of the solar spectrum. This design may lead, in future, to power levels exceeding 100 watts per sq. meter.

Research is underway to combine this novel structure with the hybrid planar-mixed molecular hetero-junctions developed in the same laboratory (Ref. 114)

The Polymer solar cells are another type of organic solar cells. They use organic compounds which are environmental friendly. They are light-weight, disposable, inexpensive to prepare, flexible and customizable at the molecular level. The efficiencies are however low. A typical view of a polymer cell is shown in Fig 2.14c.....(Ref 115)

Fig. 2.14c *Typical Polymer Solar Cell*

Polymer solar cells are the most investigated of all organic PV technologies because they are most promising.

2.6.5 III – V Solar Cells

The III – V compounds, gallium arsenide(GaAs), indium phosphide (InP) and gallium antimonide (GaSb) are compound semiconductors which have all the desirable characteristics such as direct energy band gaps, high optical absorption coefficients, good minority carrier lifetimes and mobilities (in pure, single crystal material) to become high efficient solar cell materials (Ref.78). The two materials extensively used to make single-junction solar cells are GaAs and InP with optimum band gaps, 1.42 eV and 1.34 eV respectively for energy conversion. Single-crystals of these materials are produced using either 'liquid-encapsulated Czochralski method (LEC)' or Bridgmann method (Ref. 116). The highest conversion efficiencies confirmed under standard conditions are 25.8% for GaAs, and 21.9% for InP single-junction cells, prepared with epitaxially grown homo-junction structures (Ref.117). In spite of high efficiencies, these materials are not considered as promising solar cell materials for making single-junction terrestrial solar cells, because of the very high cost to produce device-quality substrates or epitaxial layers of these compounds. The crystal imperfections including undesirable impurities severely reduce the device efficiencies; hence the low-cost deposition methods cannot be used. In addition, these materials are weaker mechanically compared to Si because these are easily cleavable. However, due to their potential for space applications, their development has been undertaken; and their very high conversion efficiencies, better power/weight ratios and better radiation resistance compared to mono-Si cells mitigate against the high material costs.

More recently, the effort has been to produce thin film multi-junction devices based on III-V materials using advanced growth methods such as 'metal organic chemical vapour deposition (MOCVD) (Ref.118). It has been shown that by increasing the number of junctions with a range of different band gap materials that cover the entire photon energies in the solar spectrum, it is possible to significantly increase the conversion efficiency. Several concepts for multi-junction cells (MJ cells) have been investigated that include mechanically stacked cells and monolithic type. In a tandem cell, by stacking solar cells of different band gaps on top of one another, with a highest band gap cell uppermost, solar light is automatically filtered as it passes through the stack. Each cell absorbs the light it can most efficiently convert, with the rest passing through to underlying cells. The monolithic type of structure has produced the highest efficiency MJ cells. The most successful materials currently used for MJ cells are the ternary III-V alloys based on GaAs and InP material systems. The development of MOCVD technique, also known as 'metal organic vapour phase epitaxy (MOVPE)' to produce GaAs- based space solar cells has led, among many developments, to the invention of high performance two-junction device (21.8% AM1.5) $GaInP_2/GaAs$ at NREL, USA in early 1980s. By 1990, the efficiency has reached 27.3% (Ref.10). The next step has been to develop triple-junction device GaInP/GaAs/Ge by Spectrolab with efficiency 37.3%,and GaInP/GaAs/GaInAs triple-junction device at 37.9% by NREL. The recent progress in the development of Ultra Triple Junction(UTJ) Space solar cells at Spectrolab has been reviewed by Yoon et al (Ref.119). Their commercial MJ devices – Advanced Triple Junction (ATJ) cells and standard triple junction devices show typical conversion efficiencies of 27.6% (AMO, 28°C) for cell areas > 26 sq.cms (Ref.120). The ATJ has a structure, InGaP/InGaAs/Ge whereas the standard triple junction has InGaP/GaAs/Ge structure. The most efficient stacked cell devices are now produced by Spectrolab in Sylmar, California (Ref.121,122). A cross-sectional view of a device with an efficiency > 39% is given in the Fig. 2.15.

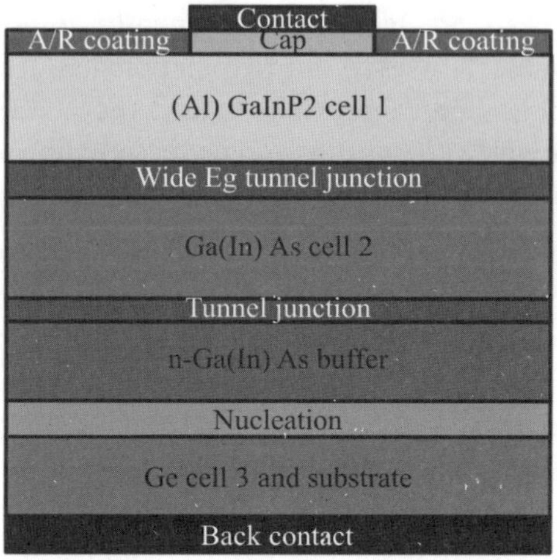

Fig. 2.15 *Cross-sectional view of a III-V multi junction device grown on a Ge subtrate with 39.3% efficiency.*

In addition to Spectrolab, Encore and Sharp companies produce commercial triple junction terrestrial concentrator cells. An overview of the research and development of MJ devices in the last 50 years , and some indication of the future direction of research is given by Sharps et al (Ref.120). Introduction of fourth-junction into the $GaInP_2$/GaAs/Ge device, viable development of mechanically stacked cells, the lattice mismatch approach to optimize the band gaps of the $GaInP_2$/GaAs/Ge device, and development of new material systems are some of the future research trends suggested in the review.

In Europe, several prominent organizations have been working on a number of approaches to achieve efficiencies between 40 and 50% (AMO) for MJ cells with increased radiation resistance (Ref.10), which include single-junction, three junction devices, and two junction concentrator cells leading to a planned six-junction device. The structure of planned six-junction cell (Ref.118) is shown in Fig. 2.16(a).

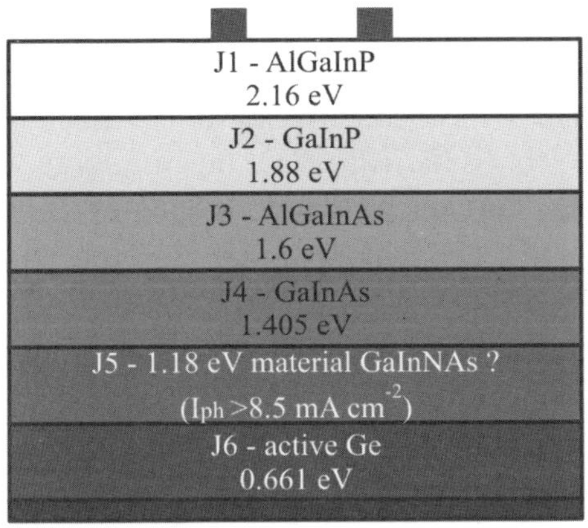

Fig. 2.16(a) *Schematic cross-sectional view of a possible 6-junction cell indicating the materials and bandgap values required. Junction 5 requires 1.18 eV material, one possible candidate is GaInAs.*

The first MJ cell with five junctions using AlGaInP, GaInP, AlGaInAs, GaInAs grown on an active Ge cell by MOVPE has been reported by Dimroath et al (Ref.123). Its schematic cross-sectional view is shown in Fig 2.16(b).This cell exhibited a Voc of 4.83 volts under three-source AM0 simulator. The same group has demonstrated a five-junction lattice-matched cell with an AM0 Voc of 5.2 volts. The efficiency, measured at 24%, is limited by the second junction at present and will be investigated by adjusting layer thicknesses (Ref.118,124). Using current simulators, it was found difficult to measure a reliable conversion efficiency.

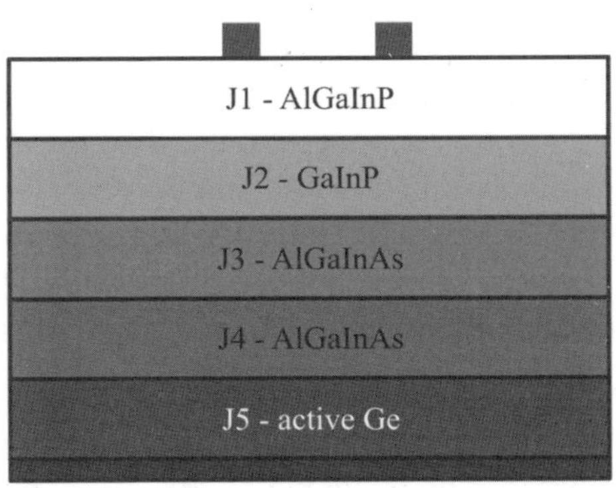

Fig. 2.16(b) *Schematic cross-sectional view of the 5-junction cell reported by Dimroth et al.*

In Japan, the Super High Efficiency Solar Cell Project, started in 1990, aims at producing a cell with conversion efficiency >40%. Development of two- and three junction devices, mechanically stacked cells, and monolithic devices are included in the project (Refs. 125-127). Japan Energy, Sumitomo Electric, NTT, and Hitachi Cable were the Japanese companies involved in this project. Sharp, the only Japanese manufacturer of MJ cells has taken over the project to develop commercial cells.

As the development of advanced semiconductor materials progress, the third-generation concepts such as intermediate band gap, and hot-carrier devices may become possible in the future. The research efforts started at UNSW by Green and his group to develop 'third generation' high performance, low cost solar cells are discussed in a series of papers (Ref. 128-131).

The high efficiencies achieved by MJ cells have reduced the effective cost per watt; however, further reduction in the cost/watt can be achieved by using low-cost substrates. The development of III-V cells on Si substrates has continued because the use of Si substrates has been identified for the next generation PV technology in USA (Ref.132a), while the use of SiGe alloys has provided a promising approach to achieve the objective.

Very recent report from NREL(Ref. 132b) shows that they have set a world record in solar cell efficiency with a PV device that converts 40.8% of the light into electricity. It is an inverted metamorphic triple junction solar cell that was designed, fabricated and measured under concentrated light of 326 suns at NREL. The new device uses compositions of Ga In P and Ga In As to split the solar spectrum into 3 equal parts that are absorbed by each of the three junctions for higher potential efficiencies. This is accomplished by growing the solar cell in a Ga As wafer, flipping it over, then removing the wafer. The resulting device is extremely thin and light and represents a new class of solar cells with better performance, design, operation and cost. Metamorphic junction are lattice - mismatched. The material properties of the mismatched semiconductors allow for greater potential conversion of sunlight.

Group III-Nitride alloys have been identified as a materials system that enables a wider variation in band gap through the variation in composition with minimum lattice mismatch. The band gap of $In_{1-x}Ga_xN$ alloys range from 0.7 eV to 3.4 eV and they have shown superior resistance to radiation compared to conventional MJ solar cell materials (Ref.133).

2.6.6 Theremo Photovoltaic (TPV) Devices

In a TPV system, appropriate PV cells are used for the conversion of radiation from radiant sources into electric power. The temperature of these radiators is around 1000 to 2000 K compared to the temperature of Sun, 6000K. Due to lack of suitable high temperature radiator materials at present, the maximum temperature of a radiator is limited to 1770K. Most of the radiation from these radiators is in infrared range. The optimum band gap of TPV device materials, therefore vary from 0.4 to 0.6 eV. These devices are used with radiators operating at temperatures, 1200 to 1500C.

TPV cells are generally located in close proximity, a few centimeters, to the source (radiator) unlike in the case of PV cells which are kept far away from the source.

The main III-V low band gap materials used are GaSb (0.72 eV), InGaAs lattice matched (0.75 eV) and mis-matched (0.5-0.6 eV) to InP, InAsSbP (0.3-0.5 eV), InGaAsSb, and InGaAs/InPAs cells. (Ref. 10). These devices have to operate at high current densities, similar to those in concentrator devices, and the open circuit voltage is generally low because of low band gaps.

To reduce current and to increase voltage, the monolithic interconnected module(MIM) structure is used which is complex and costly. The commercial TPV devices currently used are 'diffused p-n junction GaSb cells (Ref. 134). The other cells are quantum cells grown in InGaAs on InP wafers, InGaAs – based MIM, and InGaAsP/InGaAs tandem devices grown on semi-insulating InP substrates with Si3N4/Au back surface reflectors (Ref. 135-139). For a detailed review of the development and uses of TPV devices, the reader may refer to the review by Nelson (Ref.140) and the Procedings of the Thermophotovoltaic Generation of Electricity Conferences (TPV-1 to TPV-6) held during 1994-2004.

2.7 PRESENT STATUS OF DIFFERENT PV TECHNOLOGIES

Photovoltaic industry is growing faster and faster world-wide, looking at the number of emerging production sites, the expansion and creation of production capacities, the number of installations, the incentives offered by different countries and so on. The use of mono-crystalline silicon solar cells continue to grow because they are more efficient and reliable; but, the multi-crystalline silicon technology has expanded much faster in the recent years. The trend in crystalline silicon production is towards gigawatt level, higher throughput, turn key solutions from the suppliers and more automation.

The thin film technologies are booming, especially thin film silicon, and have become economically viable.

Large surface coating technologies are at operative level enabling coatings up to a substrate size of 5.7 m^2. Applied Materials (Amat) of USA and Oerlikon Solar of Switzerland have come up with turn key production lines, already operating in many locations globally. The operation of these systems are based on PECVD technique to produce a-Si and μc-Si/a-Si:H micro morph cells(and modules). This technology offers a process with low cost, stable and high production output. The production line from AMAT has an annual capacity of 40 to 80 MWp, and produce a substrate of 5.7 m^2 (2.2 m x 2.6 m). Oerlikon's production plants produce smaller modules measuring 1.4 m^2 (1.1 m x 1.3 m) and the stress is more on the quality of coating. That is why, they operate at higher plasma frequency, 40 MHz, about three times greater than that of usual PECVD in order to increase the coating quality while maintaining the high deposition rate (Ref.141,142). The core elements of the Oerlikon plants are the PECVD and TCO coating units and the laser scribing unit.

Table 2.3 shows the best efficiencies reported for different cells and modules (Ref.10).

Thin film technologies have become most advanced with regard to production technology and market penetration. The production of a-Si and μc-Si have been increasing with the availability of turn key production systems. At the end of 2007, their global production capacity has been around 500 MWp. At present the stable efficiency of commercial modules is 6 to 7%, expected to increase to 10% by 2010(Ref. 95b).

Though the process is complex and expensive, CI(G)S modules exhibit the highest production efficiencies of all thin film technologies, and the global production capacity has reached 100 MWp. The commercial modules in the market show an efficiency of 9 to12% at present (Ref. 95b)

The CdTe modules are the cheapest due to the simplicity and speed of manufacturing. The major producer, First Solar, USA alone has production capacity of 250 MWp. Their commericial modules show an efficiency of 7 to 10%, expected to go to 12% by 2010 (Ref. 95b).

The best research cell efficiencies of these technologies obtained over the period are shown in Fig 2.17a (Ref. 95a).

Table 2.3 *Best efficiencies reported for different solar cells and modules (Ref: 10)*

Type of solar cell	Highest reported small area cell efficiency			Highest reported module efficiency		
	Efficiency (%)	Area (cm^2)	Reference	Efficiency (%)	Area (cm^2)	Reference
Crystalline Si	24.7	4.0	UNSW[i], PERL[ii,2]	22.7	778	UNSW/Gochermann[3]
Multicrystalline Si	20.3	1.0	FhG-ISE[iii,4]	15.3	1017	Sandia/HEM[5]
Amorphus (and nanocrystalline) Si	10.1	1.2	Kaneka, single junction[6]	10.4	905	USSC[iv], triple junction[7]
μc-Si/αSi:H micro-morph cell	11.7	14.2	Kaneka, minimodule[8]	11.7	14.2	Kaneka, minimodule[8]
HIT[v] cell	21.8	100.4	Sanyo Corporation[9]	17.3	11 000	Sanyo Corporation[9]
GaAs cell	25.8	3.9	Kopin Corporation[10]	Not relevant	Not relevant	
InP cell	21.9	4.0	Spire Corporation[12]	Not relevant	Not relevant	
$GaInP_2$/GaAs multijunction cell	39.3*	0.4	Spectrolab, concentrator[12]	Not relevant	Not relevant	
CdTe	16.5	1.0	NREL[vi,13]	10.7	4874	BP Solarex[14]
CIGS[vii]	19.5	0.4	NREL[15]	13.4	3459	Showa Shell[16]

[i]UNSW, University of New South Wales.
[ii]PERL, passivated emitter rear locally diffused.
[iii]FhG-ISE, Fraunhofer Institute for Solar Energy Systems
[iv]USSC, United Solar Systems Corporation
[v]HIT, heterojunction with intrinsic thin layer
[vi]NREL, National Renewable Energy Laboratory
[vii]CIGS, copper indium gallium diselenide
*Boeing-Spectrolab (Sylmar, CA) announced a 40.7% efficient cell uner 240x concentrated light in December 2006 (unpublished).

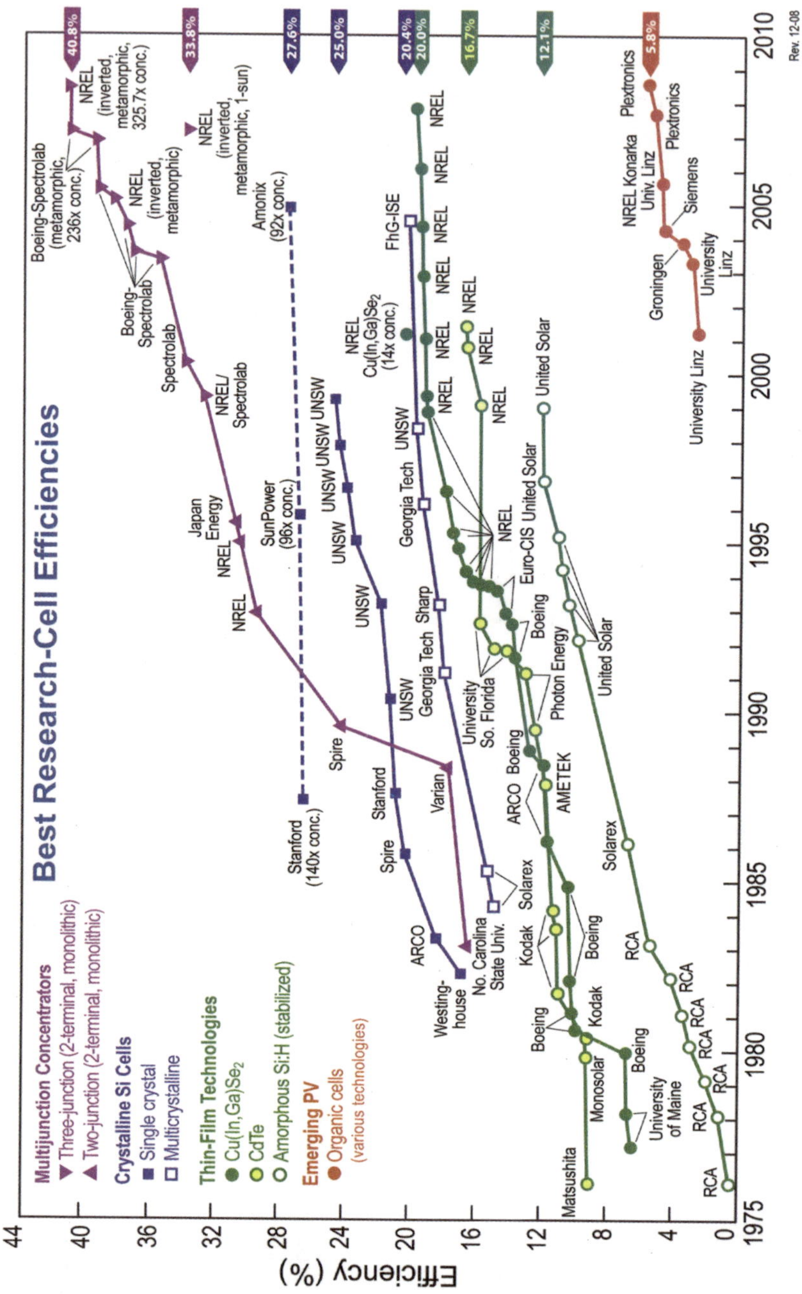

***Fig. 2.17(a)** Best Research – Cell efficiencies (Ref. 95a: NREL, 2009)*

The market share of different solar technologies sold in 2008 are shown in Fig 2.17b. The total PV shipments in 2008 are 5492 Mw (Ref. 141a).

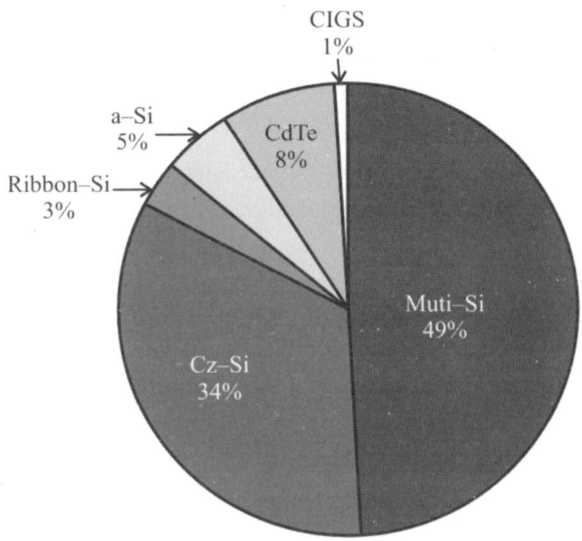

Fig. 2.17(b) *The market share of different photovoltaic solar cells during 2008 (Ref: 141a)*

The SUN &WIND ENERGY, in its report, July 2008 (Ref. 141b) listed the top 10 solar cell producers, the top 10 module producers, and the top 10 thin film module producers in 2007 and the overview for 2008. They are given below in Tables 2.4, 2.5, and 2.6. The report adopts the assessment of market research institute, EuPD Research, Bonn (Germany).

Table 2.4 *Top 10 of the cell producing companies in 2007 and 2008 (BP did not provide any data for 2008)*

2007		2008	
Sharp	710 MWp	SunTech Power	1000 MWp
SunTech Power	540	Q-cells	760
Q-cells	516	Sharp	710
Conergy	275	Gintech Energy	660
Sanyo Electrics	260	Motech Industries	580
Kyocera Solar	240	JingAo Solar	500
Motech Indust.	240	SunPower Corp.	414
SolarFun Power	240	Yingli Green	400
BP Solar	222	SolarFun Power	360
SunPower Corp.	214	Sanyo Electrics	350
		Trina Solar Energy	350

Table 2.5 *Top 10 of the module producers in 2007 and 2008 (BP did not provide any data on capacities for 2008; MSK has been bought in June 2008 by SunTech and will not be treated separately.*

2007		2008	
Sharp	716 MWp	SunTech Power	1000 MWp
SunTech Power	540	Sharp	820
Sanyo Electrics	260	Solon AG	500
SolarFun Power	240	Yingli Green	400
Kyocera Solar	240	CSI Canadian Solar	400
Solon AG	210	SolarFun Power	360
Yingli Green	200	Sanyo Electrics	350
MSK Corp.,	200	Trina Solar Energy	350
BP Solar	180.7	Kyocera Solar	300
CSI Canadian Solar	180	Schott	265

Table 2.6 *Top 10 of the thin film module producing companies in 2007 and 2008.*

2007		2008	
First Solar	308 MWp	First Solar	495 MWp
United Solar Ovonic	80	Nanosolar	220
Kaneka Solartech	75	Sharp	160
Nanosolar	60	United Solar	120
Miasole	50		
		Kaneka Solartech	90
Ersol	40	Mitsubishi Heavy	64
Schott Solar GmbH	40	Sinonar Corp.	62
Sinonar Corp.	30	Sunfilm AG	60
Honda Soltech	27.5	SunTech Power	50
Mitsubishi Heavy	24	Bangkok Solar	50

Interestingly, the new entrants from China are in the majority in top 10 silicon cell and module producers.

These assessments clearly establish that PV industry would not slow down or stagnate in the coming years, though the production targets set by different manufacturers may change due to new companies entering into market or a few others being taken over.

2.8 PRODUCTION OF SOLAR GRADE POLYCRYSTALLINE SILICON

The market share of mono- and multi-crystalline solar cells is still higher than 80%, in spite of less material consumption and low costs offered by thin film technologies. Between mono- and multi- crystalline cells, the latter is ahead in the recent years because the efficiencies are not very much different from that of former, and the production process is comparatively simple and the costs less. Fig. 2.17(b) shows the market share of different technologies in 2008.

Until 1997, the scrap/waste silicon from the microelectronics industry has been used to produce polycrystalline silicon cells. Considering the high costs involved in producing electronic-grade silicon compared to the solar-grade silicon, efforts have started to look for more economic processes to produce solar-grade silicon. The world PV has grown from 748 MWp in 2003 to 1194 MWp in 2004 (Ref.143). The demand for solar-grade silicon has been growing in the last decade, and an average annual growth of 30% is projected for the next ten years (Ref. 144). Table 2.7 shows the solar-grade poly silicon availability and demand, and projected growth for 2010 (Ref.145).

Table 2.7 *Analysis of silicon production and demand from 2003 to 2010 (in metric tons). (Ref. 145. AFB. Braga et. al, 2008).*

Year	Poly-Si demand	Available Poly-Si	Poly-Si stock
2003	9,000	9700	+700
2004	14,032	9450	-4582
2005	18,181	10115	-8066
2006	16,705	13334	-3371
2007	17,435	14979	-2456
2008	24,089	22249	-1850
2009	28,233	26973	-1260
2010	32,108	31168	-940

Lack (-) or excess (+) of poly-silicon in the worldwide market

These estimates confirm that tremendous efforts need to be focused in expanding the production plants and in finding new low-cost processes for the production of poly silicon.

The four large suppliers of solar-grade silicon since 2004 – Hemlock Semiconductors, Wacker Polysilicon, Tokuyama and Solar Grade Silicon LLC – have been using Siemens process for the purification of silicon. In this process, the trichlorosilane is fed into a reactor along with hydrogen, where thin rods of electronic-grade silicon, of about 8 to 10 mm diameter, are electrically heated to over 1100°C. Silicon from trichlorosilane or silane separates on to the hot silicon rods, reaching a diameter of 15 to 20 cms. For this to happen, it takes several days. The extracted silicon is electronic-grade, but poly-crystalline. This poly-crystalline silicon is used to grow rods of mono-crystalline silicon using Czochralski method, or ingots of multi-crystalline silicon by 'casting' process.

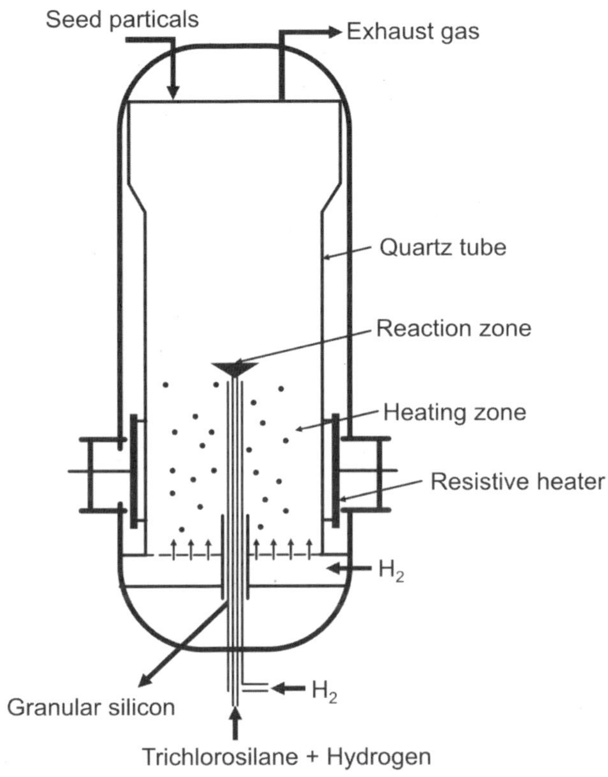

Fig. 2.17(c) *Fluidised bed reactor (Ref. 146)*

The Siemens process is highly energy-intensive and consumes about 200 kwh/kg of silicon produced (Ref.144).

Alternative manufacturing processes including metallurgical route have been under development by not only the major producers, but also the new entrants into the PV industry. Fluidised bed reactor method is one such process which has been developed to the status of pilot plant by Wacker Polysilicon, Hemlock Semiconductors Corp., and Solar Grade Silicon LLC(SGS). In this process, trichlorosilane or silane is blown into a quartz tube from below along with hydrogen. The gas stream blends with small particles of silicon which are poured into the reactor from above or which are created by the collision of large silicon granules in the reactor. Silicon separates from the trichlorosilane or silane onto the particles and thus forms larger granules. This requires process temperatures of 700°C for silane and 1000°C for trichlorosilane. At some point, the silicon grains become so heavy that they precipitate out of the gas stream onto the floor and can be removed from the reactor.

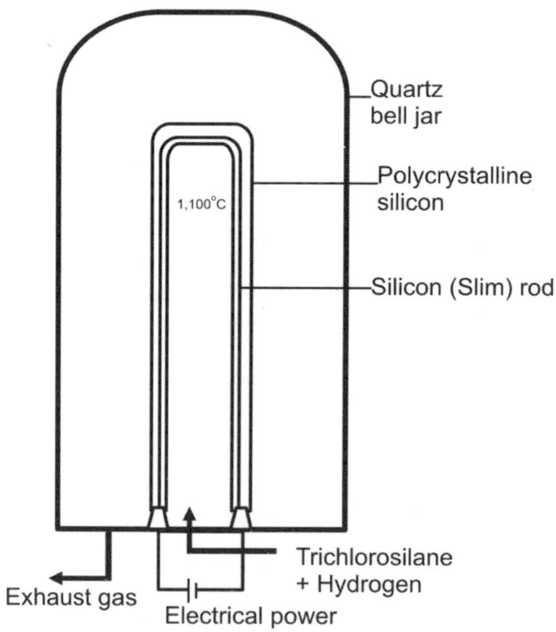

Fig. 2.17(d) *Siemens Process (Ref. 146)*

In both Siemens process and fluidized bed reactor method, the highly volatile liquid, trichlorosilane is produced by combining metallurgical grade silicon (obtained from quartz through carbothermic reaction) and hydrogen chloride. Then it is filtered from contaminants in house-sized distillation columns. The purified trichlorosilane then serves as a source material for poly-silicon.

The fluidized bed reactor method consumes significantly less energy than the Siemens process, and it can be operated continuously. The three main producers simultaneously started production of poly silicon with this alternative process also.

The metallurgical route is another process with high potential, and has been developed to the pilot plant operation stage. Here, the solar-grade silicon is obtained directly from metallurgical silicon derived from pure quartz. This route is much more energy efficient, nearly five times, than the usual Siemens process. The direct route from quartz to silicon promises a particularly low energy consumption and moderate manufacturing costs, while the feed materials are available in large quantities. PV industry estimates that annual silicon requirement would be 50,000 tons after 2010, and hence metallurgical routes are more desirable and necessary for this sort of growth.

The development status of other alternative manufacturing processes (some of them at pilot plant stage), their merits and demerits, the players involved, are covered in great detail in the latest review by AFB Braga et al (Ref. 145), and in Sun and Wind energy (Ref.146, 147).

2.9 HIGHER YIELDS FROM PV MODULES

Several PV technologies and their energy conversion efficiencies are discussed in the earlier sections. Further research and development efforts to tackle the problems identified in each of these technologies to improve upon the efficiencies and to bring down the costs are also briefly outlined. However, even with the present status of the technologies, significantly more electricity yield can be achieved by concentrating light on solar cells and/or using tracking systems.

Concentrating light on a solar cell greatly increases its output. It promises low cost and reliable solar generated power that is competitive with main-stream power generation. With the advent of multi-junction III-V solar cells, smaller and fewer cells are enough in a concentrating PV system, which enables saving of solar cell materials and bringing down the dollar-per-watt manufacturing cost compared to flat-plate modules. Tracking systems allow extracting more power from the modules because they keep the modules always normal to the direct sunlight ensuring that the modules receive maximum sunlight throughout the sunshine hours. Principles of operation of Concentrating PV systems and Tracking systems are briefly outlined.

2.9.1 Tracking Systems

A PV power generator (solar array) is positioned on a tracking system so that it remains as perpendicular to the sun as possible, as the sun moves across the sky.

Two types of systems, single-axis tracker and dual-axis tracker are used in PV installations and the tracking is performed by an electronic device. The tracking method is chosen based on astronomical data or sensor controller. Astronomical tracking systems provide a greater yield in equatorial regions with more direct sunlight than in regions with predominantly cloudy conditions. Sensor controllers, on the other hand, can turn the array to the brightest point in the sky even in cloudy conditions and improve the yield. In addition to tracking method, the increase in yield depends on geographical location, local weather and the amount of direct sun's radiation.

Tracking is effective in early morning and evening hours when the sun is low, and in locations with abundant direct sunlight. Another advantage is that tracking modules/array will be less effected by stored heat due to their distance from the ground, and thus function more efficiently due to improved operating temperatures. The inverter in the system also influences the yield; the higher yield increases are possible only with high quality inverters that have an excellent rate of efficiency in the high and partial load range (Ref. 148).

In the design of tracking systems, protection against wind and snow have to be considered. Some manufacturers – for example Lorentz, Galaxy Energy and Solon, all from Germany – have made provision to withstand the effects of wind and snow. For instance, to withstand high wind speeds, the tracker is optimized in a flow channel so that it can resist even extremely high wind speeds. When the wind speeds reach category 8 or higher, all movers automatically rotate themselves to a position with the least wind resistance. This provision helps to survive even hurricanes (Ref.148).

The most important aspect is economic viability of using tracking systems, because then help to receive 'direct solar radiation' and not the 'diffuse radiation' and 'reflected radiation'. Hence it is essential to examine whether the yield increase of 30 to 40% which the manufacturers specify can bring in more benefit than the 'additional costs' involved. With a two-axis tracking system, the 'Energy production increase' in Europe is discussed (Ref. 149).The Table 2.8 shows the observed energy production increase in different regions of Europe.

Table 2.8

Region	Energy pr3oduction increase (%)
Northern Europe (above 60 deg.)	40 - 60
Southern Europe	30 – 40
Western & Central Europe	20 – 30

At latitudes above 60 deg.N, the days of summer are very long and yet the sun is low in the sky even at midday. If the array surface is set to follow the sun, one can get a high angle for incoming radiation as well as make use of the long phases of sun rise and sunset. The energy production increase is lowest (20 to 30%) in western and central Europe and British Isles because these (mid-latitudes) lie in the frontal zone. In Southern Europe, the gain is in between.

The economic advantage is also dependent on the solar module prices. If the module prices fall and the tracking systems' costs remain the same, the economics are affected because the costs of tracking systems make up a larger share of the investment. However, tracking systems are going to dominate in the coming years in the field of concentrator PV systems because CPV requires high tracking precision. A tracker is designed to take structural flexure as well as resistance into consideration. With an accurate tracking control, it is possible to balance tolerances of structural flexure.

A list of manufacturers of single-axis and double-axis tracking systems including details of tracking technologies are given in an article by Heinz Wraneschitz (Ref.150)

2.9.2 Concentrator PV Technologies

CPV technologies have existed since the late 1970s, but they have become dominant since around 2002. The principle of CPV is quite simple. In the usual flat-plate PV array, a large area of PV material, usually crystalline silicon, is exposed to maximum light from the sun, arranging the array at an angle optimized for the latitude of the location, or by fixing on a 'tracker' which can be moved/ rotated as the sun moves across the sky. The output from the array is highest when maximum amount of solar radiation falls on it. In a concentrator system, Fresnel lenses are mostly the common solution to focus sunlight on a small area of PV material. However, other sophisticated optical concepts are also tried by a few manufacturers. If the light that falls on 100 sq.cm is focused onto 1 sq.cm of PV material, the 'concentration ratio' is considered as 100 suns;if the light from 10 sq.cm is focused onto 1 sq.cm of PV material, the ratio is 10 suns. i.e. if the concentrated sunlight falls on a CPV cell, the cell will generate at least 100 times or 10 times the electricity. At present, the commercial concentration ratios are 200 to 300 suns.

CPV system installations involve tracking systems rotating about one-axis or two-axes, to keep the solar radiation focused on the cell. With the advent of multi-junction solar cells (MJ cells), the power generated by CPV systems could be cheaper because of higher efficiencies of MJ cells. Secondly, smaller amount of PV material (cell surface area as low as 1 mm.sq.) is used compared to usual flat-plate PV systems, thereby reducing the cost. The CPV systems using smaller and fewer MJ cells are therefore highly economically viable because a lower dollar-per-watt cost is achievable, though usage of MJ cells in flat-plate modules are prohibitively expensive.

Higher concentration ratios, in practice, are limited because the ratio has to be optimized for each cell type. Beyond a certain value for the ratio, the efficiency starts to drop due to increased resistance losses within the cells. There are other issues of concern: (i) the need for accurate and reliable tracking and adequate heat sinking, and (ii) coping with environmental conditions when used in terrestrial applications.

To minimize electrical resistance, two methods are available (i) using prismatic covers that direct incoming radiation to the parts of the cell surface that lie between the metal fingers of the electrical contact grid or (ii) placing both the positive and negative electrical contacts on the back of the cell. To avoid overheating of the cells, highly conductive material such as copper is placed directly behind the cells to spread the heat or by air cooling. In practice, concentrator cells operate in the same temperature range as flat-plate PV cells.

The III-V MJ cells are capable of operating at the correspondingly high current levels, increasing from mA/sq.cm to A/sq.cm. Spectrolab of USA produces commercial triple-junction terrestrial concentrator cells with a typical 350 suns efficiency of 30.5% (10.7 W/sq.cm) and 1000 suns efficiency of 28.6% (28.6 W/sq.cm). These cells have an optimized GaInP2/GaAs/Ge structure. Sharp(Japan) and Encore (USA) also produce terrestrial MJ concentration cells (Ref.10). NREL's High Performance Project plans to achieve over 41% cell and 33% system efficiencies, with 'super cells' developed for space applications. The main driver for concentrator PV systems is 'bringing down the cost-per-watt installed or per kWh produced.'

Jackie Jones has outlined details of CPV developed/offered by various companies worldwide (Ref.151). A global survey recently carried out on Concentrating PV systems by Sun & Wind Energy (Ref.152) lists the various companies along with the development activities and product specifications. A few optical designs under study for concentrating sunlight, are also discussed.

2.10 PERFORMANCE OF SOLAR CELL

The sun is an immense source of power, emitting light and energy over a range of wavelengths. The amount of solar radiation reaching the earth's surface varies greatly due to changing atmospheric conditions and the changing position of the sun during the day and the year. The clouds, aerosols, other particles in the atmosphere reflect, absorb and scatter the solar radiation and part of it only reaches the earth's surface directly.

The local geographical features, such as mountains, large water bodies, coastlines and plains influence the atmospheric conditions which affect the solar radiation reaching earth's surface. The total radiation at any point on earth's surface is referred as 'global solar radiation' and consists of 'direct' radiation and 'diffuse' radiation. The direct radiation comes in a direct line from the sun, whereas 'diffuse' radiation is the scattered radiation by the dust, water molecules, clouds and aerosols in the atmosphere. This is also called sky radiation. The diffuse radiation is about 10 to 20% for clear skies and may go up to 100% for cloudy skies.

The amount of solar radiation received at noon is maximum because when the sun is directly overhead (zenith position), the thickness of the atmosphere is minimum at sea level and is referred as 'air mass 1, (AM1) The solar energy available at earth's surface under this condition is 1000 watts/sq.meter for a clear sky (Fig. 2.18). Any change in the sun's angle (air mass becoming greater than 1) will reduce the solar energy available on the earth's surface. The air-mass value at any time of the day depends on the latitude of the place, the time of the day and the day of the year. When the Sun is overhead, it is lower in the sky.

When Sun is closer to horizon, the 'direct radiation' travel longer distances through the atmosphere resulting in more absorption and scattering of the radiation. The effect of the atmosphere is precisely to create spatial, temporal and spectral variation in solar radiation which influence the solar energy conversion systems on the earth's surface. A brief description of the instruments used to measure the direct, diffuse and total/global solar radiation are given in Annexure 2. The spectral distribution of solar radiation is given in Fig. 2.19. The spectral distribution outside the atmosphere is significantly different from that on earth's surface, due to absorption and scattering of radiation by the atmosphere.

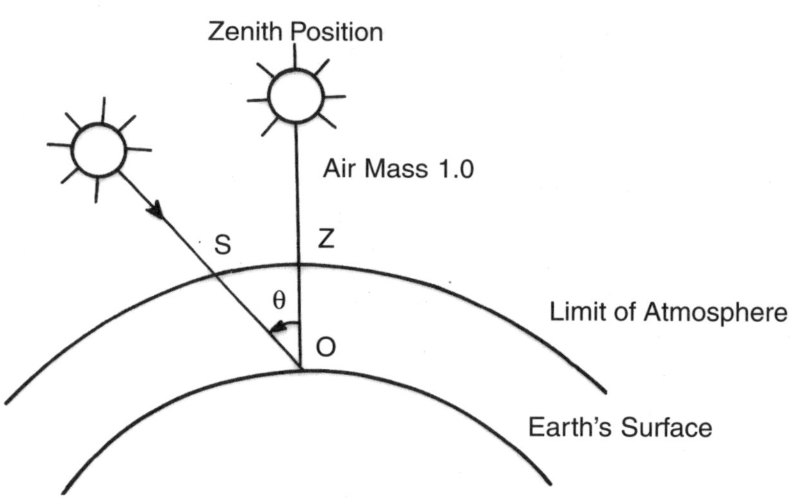

Fig. 2.18 *The elevation of Sun above the horizon. When the Sun is at an angle θ, to the zenith, the air mass is SO/ZO =* ***Sec*** $\boldsymbol{\theta}$

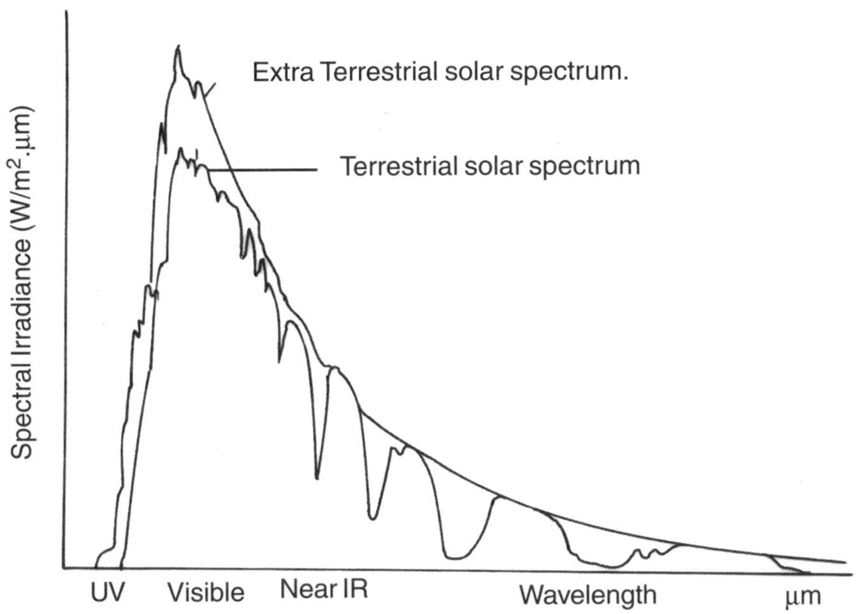

Fig. 2.19 *Spectral Distribution of Solar Radiation*

Also, the location (latitude), climate and atmospheric conditions influence the spectral distribution throughout the day and the year, affecting the percentage of energy in UV, visible and near IR regions. The study of the spectral distribution during the day and the year are highly essential because solar cells respond primarily to the light photons in the visible and near-IR regions of the spectrum.

Consider the solar radiation wavelength sensitivity of silicon solar cell shown in Fig. 2.20. The photons with short wavelengths (i.e., high energy ones) are immediately absorbed as soon as they incident on the front surface of the cell; and if they are absorbed far away from the junction, they will not effectively contribute to the electron current, i.e., some of the high energy photons are ineffective in generating a current.

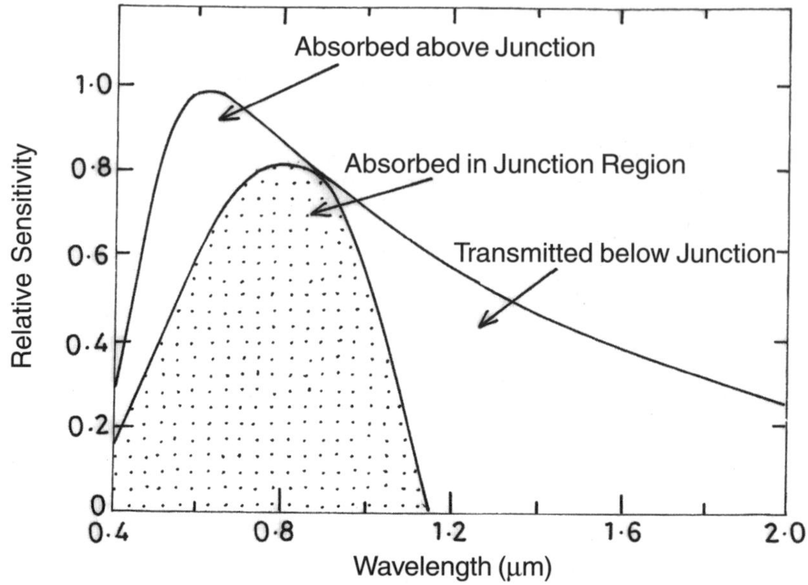

Fig. 2.20 *Solar Radiation wavelength sensitivity of a silicon solar cell*

The long wavelength photons are transmitted through the junction and are either absorbed in the bottom of the cell or transmitted through the cell producing heat. The photons in the medium wavelength range around 0.8 micrometers, are absorbed near the junction and effectively contribute to the current. It means, the spectral energy equal to E_g only produces charge carriers generating electric current, and the rest of the spectrum contributes to generation of heat.

For crystalline silicon solar cell, about 20% of solar energy is less than E_g and is not absorbed by the cell; 30% of energy is greater than E_g which dissipates as heat; and 20% accounts for the incident losses in the cell due to reflection at the surface, metallization and internal resistance of the cell. As a result, only 30% of solar energy is available for conversion into electric energy. It is therefore not surprising that the theoretical efficiency of silicon is about 20% only. The inherent losses could partly be reduced by engineering the material.

An antireflection coating would reduce surface reflections to a few percent facilitating more than 90% of light energy to enter into the cell. Surface reflections could be further reduced by the use of multilayer coatings with different refractive indices. The electrical performance of a solar cell is evaluated by studying the I-V relationship. The current, I is proportional to sun light intensity whereas the voltage, V is characteristic of the semiconducting material. The I-V curve relates the cell's output power response to an external load that varies from zero-load level (short-circuit condition) to 'infinite-load' level (Open-circuit condition).

The I-V curves of a silicon solar cell at different light intensities are shown in Fig.2.21. One sun is equivalent to 1000 watts/square meter. The power output of the cell is almost directly proportional to the intensity of sunlight.

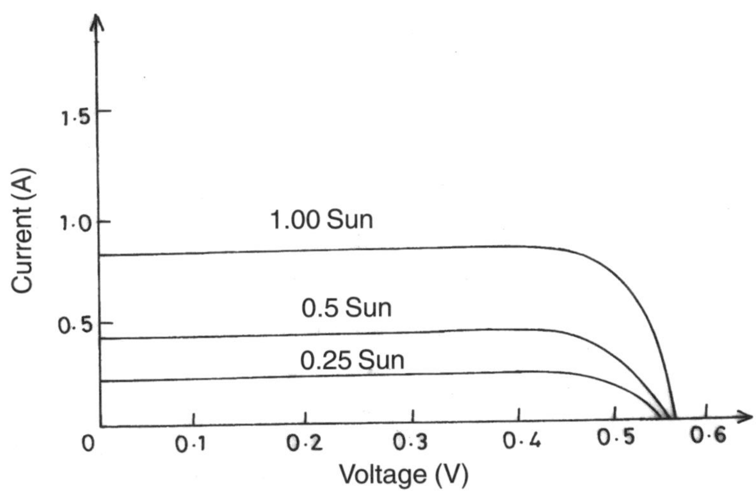

Fig. 2.21 *I-V curves of silicon solar cell at different light intensities*

The important feature is that the voltage of the cell does not depend on the size and light intensity and remains constant at around 0.5V dc. But the current in the cell changes with the size and light intensity. To compare the performance of cells of different sizes, a parameter called 'current density 'is used as an index.

2.10.1 Spectral Response

The spectral response (SR) of a cell is defined as the ratio of the number of charge carriers collected by the junction (ie, photocurrent) to the intensity of the incident light at a given frequency/ wavelength. The major contribution to the 'spectral response' and photocurrent come from the base material in addition to a small contribution from the depletion region. The nature and quality of the cell are known by studying the spectral response.

It is expressed as where I_{ph} is the photocurrent and $E(\lambda)$ is the energy of the incident photon of wavelength,

$$SR(\lambda) = I_{ph}(\lambda)/ q.E(\lambda) \qquad \text{........ (1)}$$

The response generally begins at the band gap energy, reaches a peak and then decreases due to recombination in the diffused region. The ideal hetero-junction cell gives high and constant spectral response between the band gaps of window and absorber layer materials.

The shape of SR is influenced by the quality of the two materials and the interface. If the diffusion length in the absorber is low, the carriers produced are lost through recombination, reducing the red response of the solar cell. Impurities in the window layer may create traps within the band gap leading to absorption at wavelengths above the absorption edge which tends to reduce the blue response of the cell. The effect of temperature, crystal defects and the internal resistance of the cell on its performance /efficiency are briefly mentioned.

2.10.2 Effect of Temperature

The changes in ambient temperature influences the performance of the solar cell. The efficiency of the cell gets reduced with the increase of cell temperature. V_{oc} is sensitive to temperature whereas Isc is not. Simple calculation may show that the cell voltage and temperature are inversely related (Ref. 153)

$$dV_{oc}/dT = (1/q).(dE_g/dT) - (1/T).[(E_g/q) - V_{oc}] \qquad \text{........ (2)}$$

For silicon cell, (dV_{oc}/dT) is approximately equal to –2 mV/°C, which means, that the efficiency of the cell drops by about 0.4 % for increase of every one degree Celsius. A silicon solar cell of 20% efficiency at 20°C will reduce to 16% at 30°C.

2.10.3 Effect of Electronic Defects

The electron-hole recombination is one of the major factors that affect the efficiency of the cell. Larger recombination lifetimes are desirable for ensuring high I_{sc}. The material processing is very important,- otherwise the presence of electrically active point defects, crystal dislocations and grain boundaries promote early recombination of electron-hole pairs, lowering recombination life times. The extended lattice defects in the cell may cause shunt resistance which also affects the performance of the cell.

2.10.4 Effect of Metallization and Cell Resistance

The series resistance of a cell arises from the contact leads, the metal contact grid and the bulk resistance of the cell. The shunt paths across the junction also result in power loss. The series resistance of the top layer can be minimized by spacing metal lines closely.

Generally, the effect of finite series resistance on the efficiency is minimal unless the material is amorphous or an organic semiconductor with very high resistivity. For a typical silicon solar cell, the series resistance is about 0.7 ohm, and it should reduce to zero for an ideal cell. The substrate resistivity is also an important factor because it influences both V_{oc} and I_{sc}. For silicon cells, the highest efficiencies are observed for substrate resistivities between 0.1 and 1.0 ohm-cm (Ref. 154).

REFERENCES

1. ***D.M.Chapin et al:*** *J. Appl. Phys.* ***25****, p676 (1954)*
2. ***D.C.Reynolds et al:*** *Phys. Rev.* ***96****, p533 (1954)*
3. ***SARASIN BASIC REPORT:*** *PV2001 - The Market Players and Forecasts', (2001)*
4. ***J.P.Benner et al:*** *IEEE Spectrum,* ***36****, p28 (1999)*
5. ***M.A. Green et al*** *: Prog. Photovolt, Res. and Applications, p1349 (2005)*
6. ***M.A. Green:*** *Silicon Solar Cells- Advanced Principles and Practices, Bridge Printery, Sydney (1995)*
7. ***J.F. Nijis et al:*** *Proc.16th European PV Solar Energy Conf. p1064 (2000)*

8. ***G.P.Willeke***: *Proc.19th European PV Solar Energy Conf.,Paris, p383 (2004)*

9. ***R.W.Miles et al***: *Vacuum, **80**, p1090 (2006)*

10. ***R.W.Miles et al***: *Materials Today, **10**, p20 (2007)*

11. ***M.A.Green***: *Tech.Digest, Intl.PVSEC-14, Bangkok, p391 (2004)*

12. ***S.Wenham***: *Prog. Photovolt. Res. Applications, **1**, p3 (1993)*

13. ***A.Mason et al***: *Proc.19th European PV Solar Energy Conf.,Paris, p2653 (2004)*

14. ***J.Zhao et al***: *Appl.Phys.Letts, **66**, p3628 (1995)*

15. ***Zhao et al***: *Appl.Phys.Letts, **73**, p1991 (1998)*

16. ***T.Baba et al***: *Proc.13th European PV Solar Energy Conf., p1708 (1995)*

17. ***M.Ghosh et al***: *Proc.19th European PV Solar Energy Conf,Paris, p560 (2004) W. Kock et al: Porc. 11th PVSEC, Montreux p518 (1992)*

18. ***A.Eyer et al***: *Optoelectronics, **5**,p239 (1990)*

19. ***M.D.Rosenblum et al***: *Proc.29th IEEE PV Specialists Conf, New Orleans, p58 (2002)*

20. ***Sun & Wind Energy***: *Draw or Saw – a system comparison, p111 (2/2006)*

21. ***W.Schmidt et al***: *Proc.16th European PV Solar Energy Conf. p1082 (2000)*

22. ***R.L.Wallace et al***: *Solar Energy materials & Solar Cells, **48**, p179 (1997)*

23. ***Sun & Wind Energy***: *Cells out of the ribbon, p55 (2/2004)*

24. ***C.R.Wronski et al***: *Amorphous Si solar cells, In: M.D. Archer, R.Hill eds. Clean electricity* from Photovoltaics, Imperial college Press, p199 *(2001)*

25. ***S.Amiruddin et al***: *Tech.Digest, Intl.PVSEC-14, Bangkok, p111 (2004)*

26. a. ***D.E.Carlson et al***: *Appl. Phys. Letts,**28**, p671 (1976)*

    b. ***S.Guha***: *Tech.Digest, Intl.PVSEC-14, Bangkok, p33 (2004)*

    c. ***Deng. x. et al***: *In handbook of PV Science and Engineering, A. Luque and S.Hegedus (Eds), John Wileys, p505-566 (2003)*

    d. ***AV. Shah et al***: *Prog. in photovoltaics, **12**(2-3), p113-142 (2004)*

    e. ***J. Meier et al***: *Solar energy, 77, p983 (2004)*

27. ***K.Kawamati et al***: *Tech.Digest, Intl. PVSEC-14, Bangkok, p41 (2004)*

28. ***E.Maruyama et al***: *Proc.4th World Conf. on PV Energy Conversion, p1455 (2006)*

29. ***R.W. Miles et al***: *Progr. Crystal growth and characterisation of materials, **51**, p1 (2005) Y.Hamakawa: Tech.Digest, Intl. PVSEC-14, Bangkok, p1 (2004)*

30. ***H.W.Schock et al***: *Renewable Energy World, 4, p75 (2001)*

31. a. ***K.Kishimoto et al***: *Parc. 20th European PVSEC, Barcelona, (2005)*

    b. ***K.Yamamoto et al***: *Prog. in photovoltaics, **13**, p489 (2005)*

32. a. ***R.B.Bergmann et al***: *Proc.Photovolt.Res.Applications, 8, p451 (2000)*

    b. ***A.Shah et al***: *Tech.Digest, Intl. PVSEC-14, Bangkok, p39 (2004)*

33. a. ***A.Catalano***: *Solar Energy Mater. Sol.Cells, **41/42**, p205 (1996)*

    b. ***Z. Shi et al***: *Prog. Photovoltaic Res. Appl., **6**, p247 (1998)*

    c. ***J.Zhao et al***: *Prog. Photovoltaic Res.Appl., 7, p471 (1999)*

34. ***R.B.Bergmann:*** *Appl. Phys.,* ***A69****, p187 (1999)*

35. ***M.J.Mc Cann et al:*** *Solar Energy Mater. Sol. Cells,* ***68****,p135 (2001)*

36. ***K.R.Catchpole et al:*** *Solar Energy Mater. Sol. Cells,* ***68****, p173 (2001)*

37. ***R.Brendel:*** *Japan J. Appl. Phys.,* ***40****, p4431 (2001)*

38. ***R.B.Bergmann et al:*** *(a)Thin Solid Films,* ***403-404****, p162 (2002) (b)Solar Energy Mater. Sol. Cells, (2002)*

39. ***H.Tayanaka et al:*** *(a) Proc.6th Sony Research Forum, p556 (1996) (b)2nd World Conf. on PV Solar Energy Conversion, Vienna, Ed by J Schmidt et al, p1272 (1998)*

40. ***R.Brendel:*** *14th European PV Solar Energy Conference, Ed. by H.A.Ossenbrisk et al, p1354 (1997)*

41. ***K.J.Weber et al:*** *Appl. Phys.,* ***A69****, p195 (1999)*

42. ***S.Oelting et al:*** *12th European PV Solar Energy Conference, Ed by R.Hill et al, p1815 (1994)*

43. ***T. Matsushita et al:*** *16th European PV Solar Energy Conference., p1679 (2000)*

44. ***S.Wagner et al****: Appl.Phys.Letts.,****25****, p434 (1974)*

45. ***T.Dullweber et al****: Solar Energy materials & Solar cells,* ***67****, p145 (2001)*

46. ***L.L.Kazmerski et al****: Appl.Phys.Letts.,* ***46****, p268 (1976)*

47. ***K.Ramanathan et al****: Prog.Photovoltaic Res. Appl.* ***11****, p225 (2003)*

48. ***K.Siemer et al****: Proc.16th European PV Solar Energy Conf, p895 (2000)*

49. ***M.Powalla et al****: Thin Solid Films,* ***361-362****, p540 (2000)*

50. ***B. Dimmler et al:*** *Proc. 31st IEEE PV specialist conference, p189 (2005)*

51. ***B.M.Basol et al:*** *IEEE Transactions on Electronic Devices,**37**, p418 (1990)*

52. ***R. Weiting et al:*** *Proc. 31st IEEE PV Specialist conference, p177 (2005)*

53. ***M. J. Carter et al:*** *Proc. 19th IEEE PV Specialist conference, New Olleans, p1275 (1987)*

54. ***U.Rau and H.W.Schock:*** *In M.D.Archer and R.Hill eds. Clean electricity from Photovoltaics, Imperial college Press p277 (2001)*

55. ***A.Romeo et al:*** *Prog.Photovoltaic Res.Appl.,**12**, 93 (2004)*

56. ***D.Rudmann et al:*** *Thin Solid Films,**480-481**, p55, (2005)*

57. ***D.Rudmann et al:*** *Appl.Phys.Letts.,**84**,p1129 (2004)*

58. ***S.Siebentritt:*** *Solar Energy,**77**, p767 (2004)*

59. ***T.Minemoto et al:*** *Thin Solid Films,**372**,p173 (2000)*

60. ***J.Kessler et al:*** *Proc.16th European PV Solar Energy Conf. (2000)*

61. ***H.S.Ullal et al:*** *Proc.29th IEEE PV Specialists Conf.,New Orleans,p472 (2002)*

62. ***H.S.Ullal et al:*** *Proc.19th European PV Energy Conf., p1678 (2004)*

63. ***W.N. Shafarman et al:*** *In Handbook of Photovoltaic Science and Engineering, A. Luque and S.Hegedus (eds), John Wileys, p567–616 (2003)*

64. ***A.Romeo et al:*** *Prog. in Photovoltaics, **12**, p93 (2004)*

65. ***M.A.Contreras et al:*** *Prog.Photovoltaic Res. Appl. **13**, p29 (2004)*

66. ***Y.Tanaka et al:*** *Proc.17th European PV Solar Energy Conf.,Munich,p989 (2001)*

67. ***M. Powalla et al:*** *Proc. 16th European PV Solar Energy Conf., p275 (2000)*

68. ***M. Powalla et al****: Thin Solid Films,* ***387****, p251 (2001)*

69. **a.** ***M.Powalla et al:*** *Tech Dig. Intl. PVSEC - 14, Bangkok, p513 (2004)*

    b. ***B. Dimmler:*** *Proc. 23rd European PVSE conference, Valencia (2008) B. Dimmler: Low cost Solar energy, Savilla, Feb (2009)*

70. ***T.Nagami et al:*** *Thin Solid Films,* ***403-404****, p197 (2002)*

71. ***W.Stetter et al:*** *Proc.16th European PV Solar Energy Conf., p783 (2000)*

72. ***D.Rudmann et al:*** *Proc.16th European PV Solar Energy Conf., p298 (2000)*

73. ***R.Klenk et al:*** *Proc.16th European PV Solar Energy Conference., p325 (2000)*

74. ***I. Forbes et al:*** *Proc. 18th European PVSE conf., p 1939 (2004)*

75. **S. Armstrong *et al*:** *Thin Solid Films, 403-404, p126 (2002)*

76. **a.** ***K.T.R. Reddy et al:*** *Tech. Digest, PVSEC -14, Bangkok, p699 (2004)*

    b. ***R.Scheer et al*** *: Solar Energy, 77, p777 (2004)*

    c. ***H.J.Lewerenz*** *: Solar Energy Materials and Solar cells, 83, p395 (2004)*

    d. ***R.Klenk et al*** *: Thin Solid Films, 480-481, p509 (2005)*

    e. ***Jo Klaer et al.*** *: Thin Solid Films, 515, p5929 (2007)*

    f. ***B.Berenguir et al*** *: Electrochemistry Communications, 8, p165 (2006)*

    g. ***T. Theresa John*** *: Solar Energy materials and Solar cells, 89, p27 (2005)*

    h. ***B. Asenjo et al*** *: Thin Solid Films, 515, p6036 (2007)*

i. ***B. Asenjo et al*** : *Solar Energy Materials ans Solar cells, 92, p302 (2008)*

j. ***S. Nakamura*** : *Solar Energy materials and Solar cells, 75, p81 (2003)*

77. ***J.J.Loferski***: *Proc.12th IEEE PV Specialists Conf. p853 (1976)*

78. ***N.A.Goken et al***: *Solar Energy materials,**1**,271 (1979)*

79. ***R.H.Bube et al***: *Fundamentals of Solar Cells, Academic Press, NY (1983)*

80. ***R.H.Bube***: *Photovoltaic materials, Imperial College Press, London (1998)*

81. ***K.Zweibel et al***: *J.Soc.Photo-optic Instr.Engg. **543**, p119 (1985)*

82. ***D.Bonnet et al***: *Proc.9th IEEE PV Specialists Conf. Silver Springs, p129 (1972)*

83. ***X.Wu et al***: *Proc.16th European PV Solar Energy Conf. p341 (2000)*

84. a. ***X.Wu et al***: *Proc.17th European PV Solar Energy Conf,.Munich,p995 (2001)*

b. ***Aramoto. T. et al:*** *Solar Energy Materials and Solar cells, **75**, p211 (2003)*

85. ***D.Bonnet***: *Thin Solid Films, **361-362**, p547 (2000) & Proc.19th European PV Solar Energy Conf, Paris, p1657 (2004)*

86. ***T.Takamoto et al***: *Solar energy mater. & Solar Cells, **49**, p219 (1997)*

87. ***C.C.Morris et al***: *Proc.21st IEEE PV Specialists Conf, p579 (1990)*

88. a. ***T.L.Chu et al***: *Solid State Electronics, **38**, p533 (1995)*

b. ***B. Mc Candless et al:*** *In Handbook of PV Science and Engineering, A. Luque and S. Hegedus (Eds), John Wileys, p617 (2003)*

89. ***G.Zoppi et al:*** *Semicond.Sci.Technol.* ***21****, p763 (2006)*

90. ***A.Romeo et al:*** *Proc.2nd World Conf. on PV Solar Energy Conversion, Vienna, p1105 (1998)*

91. ***X.Wu et al:*** *Proc.28th IEEE Photovoltaic Spl. Conf, p470(2000)*

92. ***H.S.Ullal et al:*** *Proc. 14th IEEE PV Sci. and Engg. Conf., p417 (2004)*

93. ***A.Abken et al:*** *Proc.16th European PV Solar Conf, p820 (2000)*

94. ***K.L.Barth et al:*** *Proc.29th IEEE Photovoltaic Spl. Conf.,(2002)*

95. a. ***Private communication from H. Ullal, NREL, Jan 2009***

    b. ***Sun & Wind Energy****: Thin film is self-assured, p120-123, (2/2008)*

96 a. ***D.Eaglesham:*** *Thinfilm compound Semiconductor PV Symposium, Materials Research Society, Spring meeting (2007)*

   b. ***News item from First Solar,*** *24th Feb (2009)*

97. ***H.S.Ullal:*** *Tech.Digest.Intl.PVSEC-14, Bangkok, p417 (2004)*

98. ***H. Tributsch et al:*** *Ber.Bunsenges.Phys.Chem.,72, p437 (1968)*

99. ***B.O'Reagan et al:*** *Nature,* ***353****, p737 (1991)*

100. ***A.Hinsch et al:*** *Proc.17th European PV Solar Energy Conf. Munich, p995 (2001)*

101. ***M. Gratzel:*** *Photo electrochemical cells, Nature,* ***414****, p338 (2001)*

102. ***M.Graetzel:*** *Prog. Photovoltaic Res. Appl.,* ***8****, p171 (2000)*

103. ***J. He et al:*** *Solar Energy Materials and Solar cells,* ***62****, p205 (2000)*

104. ***M. Durr et al:*** *Proc. 19th European PVSEC, Paris, p21 (2004)*

105. ***O. Chevalevski et al:*** *Proc. 3rd world conf. PVEC, Osaka, p23 (2003)*

106. ***W. Kubo et al:*** *J. Photochem. Photobiology A,* ***164****, p33 (2004)*

107. ***P. Wang et al:*** *J. American chemical Society,* ***127****, p808 (2005)*

108. ***M. Gretzel:*** *Chemistry Letters,* ***34****, p8-13 (2005)*

109. ***M.Graetzel:*** *MRS Bulletin,* ***30****, p23 (2005)*

110. ***M. Graetzel:*** *Prog. Photovolt. Res. Apply.,* ***14****, p429 (2006)*

111. ***Schulze et al:*** *Advanced Materials,* ***18****(21), p2872 (2006)*

112. ***www.iapp.de/iapp/agruppen/osol/research:organic_solar_cells:basics_of_osc.14k***

113. ***B.P.R and et al:*** *Appl. Phys. Letts,* ***87****, 233508, p233508-1 to 3(2005) A.L. Wadell et al: Article in Greenpower Magazine, (2005)*

114. ***J. Xne et al:*** *J. Appl. Phys.* ***98****, 124903, pp 124903 - 1 to 9 (2005) B.P.R and et al: J. Appl. Phys.,* ***98****, 124902, p12492- 1 to 7 (2005)*

115. *http//:wikimeida.org/wikipedia/en/8/87/solar cells 4.gif*

116. ***M.E.Klausmeier-Brown*** *:Status, Prospects and Economics of Terrestrial, Single junction GaAs concentrator cells' in 'solar cells and their Applications' Partain,L.D (ed), John Wileys, p125 (1995)*

117. ***M.A.Green et al:*** *Prog.Photovolt.Res.Appl.,****13****, p387 (2005)*

118. ***M.Meusel et al:*** *Proc.19th European PV Solar Energy Conf., Paris, p3581 (2004)*

119. ***H.Yoon et al:*** *Prog.Photovolt.Res.Appl.,****13****, p133 (2005)*

120. ***P.R.Sharps et al:*** *Proc.19th European PV Solar Energy Conf.,Paris,p3569 (2004)*

121. ***R.R.King et al:*** *Proc.21st European PV Solar Energy Conf.,Dresden,p124 (2006)*

122. ***M.A.Green et al:*** *Prog.Photovolt.Res.Appl,* ***15****, p35, (2007)*

123. ***F.Dimroath et al***: *Proc.Third World Conf. on PV Energy conversion (2003)30-D9-01*

124. ***N.Pearsall***: *Renewable Energy World, **7**(4), p247 (2004)*

125. ***M.Yamaguchi et al***: *Prog.Photovolt.Res.Appl,**13**, p125 (2005)*

126. ***M.Yamaguchi et al***:*Solar Energy Materials & Solar Cells, **79**, p78 (2005)*

127. ***M.Yamaguchi et al***:*Solar Energy Materials & Solar Cells, **75**, p261 (2003)*

128. ***G.J.Conibeer et al***:*Proc.19th European PV Solar Energy Conf, Paris, p270 (2004)*

129. ***M.A.Green***: *Proc.Third World Conf. on PV Energy Conversion (2003)10-B13-01*

130. ***M.A.Green***: *Third Generation Solar-future PV, Renewable Energy World, **7**(4), p203 (2004)*

131. ***M.A.Green***: *Proc. 19th European PVSEC, Paris, p3-8 (2004)*

132. **a.** ***R.McConnell et al***: *Proc.19th European PV Solar Energy Conf, Paris,p92 (2004)*

132. **b**. *NREL News Release, NR-2708, August 2008*

133. ***W. Walukiewicz et al***: *Proc.19th European PV Solar Energy Conf, Paris, p30 (2004)*

134. ***L.M.Frass et al***: *Proc. 4th NREL Conf. on TPV Generation of electricity(TPV-4), p480 (1999)*

135. ***C.Rohr et at***: *Proc. 5th Conf. TPV Gen. of electricity(TPV-5), p 344 (2002)*

136. ***D.M.Wilt et al***: *Proc. 1st NREL Conf. on TPV Gen. of electricity(TPV-1), p210 (1994)*

137. ***S.L.Murray et al***: *Proc. 19th European PVSEC, p97 (2004)*

138. ***J.S.Ward et al***: *Porc. 3rd NREL Conf. on TPV Gen. electricity, p227 (1997)*

139. ***R.R.Siergieg et al***: *Proc. 6th Conf. on TPV Gen. electircity (TPV-6), p480 (2004)*

140. ***R.E.Nelson***: *Semiconductor Science and Technology,* ***18*** *p141 (2003)*

141. **a**. *Private Communication from Northumbria University (Prof. Pearsall laboratory), Jan 2009*

141. **b.** ***Sun & Wind Energy***:*Oerlikon: their grip on the market, p168 (4/2008)*

142. ***Sun & Wind Energy***:*The technology: a search for the ideal approach, p125 (1/2008)*

143. ***Private communication from H.Ullal:*** *NREL (2009)*

144. ***A.Wearness***: *Solar grade silicon by direct metallurgical process,SINTEF/ECN/SCANARC/SUNERGY, Silicon for chemical industryIII July (2006)*

145. ***A.F.B.Braga et al***: *Solar Energy materials & Solar Cells,* ***92****, p418 (2008)*

146. ***Sun & Wind Energy***: *Solar grade silicon: expensive and in short-supply, p78-82 (1/2005)*

147. ***Sun & Wind Energy***: *Full stream ahead from 2008, p418-424 (2/2006)*

148. ***Sun & Wind Energy***: *Tracking for More Profit, International Issue, p28 (2/2006)*

149. ***Sun & Wind Energy***: *Tracking Systems, Higher Output, Higher Investement Castes, Intl. Issue p136 (4/2008)*

150. ***Sun & Wind Energy***: *Racing after the Sun, Intl. Issue p18(2/2006)*

151. ***Jackie Jones***: *Renewable Energy World,* ***8****(5), p78 (2005)*

152. ***Solar & Wind Energy:*** *Concentrator Systems – Sophisticated Optics, p 148 (3/2007)*

153. ***R.S.Muller et al:*** *Device Electronics for Integrated Circuits, Wiley-Inter Science, New York (1977)*

154. ***A.E.Dixon:*** *Solar Energy Conversion-II, Edited by A.F.Janzen and R.K.Swartman, Pergmon Press, Toronto (1981)*

CHAPTER THREE

SOLAR MODULES AND ARRAYS

3

SOLAR MODULES & ARRAYS

3.1 INTRODUCTION

A single solar cell produces a voltage of about 0.5 volts regardless of its size. If the cells are connected in series, the voltages from the cells are added at a constant current level; and if they are connected in parallel, the currents get added at constant voltage level.

Generally cells are interconnected in series to obtain a configuration that decides the required voltage and current. An environmentally protected assembly of solar cells designed to generate required electric power at a particular current rating and voltage is called a solar 'module' or 'panel'. The module normally operates at 12 volts dc and at a wattage of a few watts to about 100 to 250 Watts (Fig. 3.1). There is a problem in connecting cells together. Since the current generated is proportional to cell area, any loss of area due to shade or physical crack in the cell, results in less than the normal current output. Such a cell acts like a power dissipator and not as a power generator, because the cell operates in the negative voltage region of its I-V curve. This problem is overcome by connecting bypass diode around the cell,- if a cell is shaded and thus turned off, the current will bypass that cell so that the module is not totally shut down.

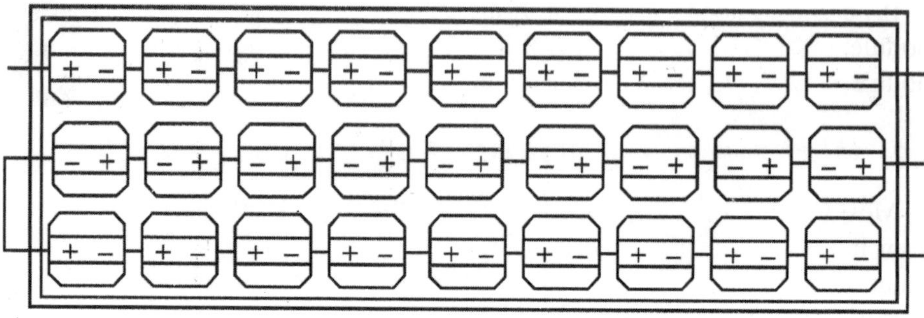

Fig. 3.1 *PV module with 36 cells in series*

The modules are made in a wide range of sizes to suit different applications:

(i) *low voltage/low power modules for voltages less than 6 volts and out put power of a few milli watts for application in watches, calculators, cameras and other devices that operate on small power. Generally, a-silicon is used to make these low power modules,*

(ii) *Modules of 1 to 10 watts at 3 to 12 volts for use in toys, radios, electric fences, lanterns, small pumps etc. These are made from broken pieces of single crystal or polycrystalline cells or from a-silicon sheets,*

(iii) *Large modules of 10 to 250 watts at 12 or 24 volts. These are made from full sized cells, connecting them in series, the configuration changing depending on the output power required. The modules are used in lighting, communications, water pumping, domestic solar power units, large size solar power plants and many other applications. High power modules are generally made of single crystal silicon cells; and with rapid developments in the manufacturing of poly- and amorphous silicon materials, they are also used in the fabrication of high power modules.*

If an application demands more power than can be provided by a single module, large units called 'arrays' are assembled by connecting a number of modules,- the number varying from tens to hundreds.

3.2 MODULE SPECIFICATIONS

Most solar modules are made for 12 volt applications. Efficiency of a module is an important specification, which is indicated by how much of the solar radiation falling on the module is converted into electric power. If the sun light reaching the earth at sea level on a bright clear day is about 1000 watts/sq.m, a module with 12% efficiency will convert that solar energy to 120 watts of electric power, while the rest converts into heat. If due to a heavy overcast sky, only 5% of normal solar light is available, one can get 5% of the maximum output power from the module. The efficiency of a module is not sum total of the efficiencies of individual cells used in the module due to the losses from wiring, backing and frame.

The other specifications of a module are specified at 'standard test conditions' (STC) at 25°C. Most of the modules on the field are exposed to higher temperatures, and deliver less output power. The 'specifications' of a module/panel are the following:

(i) ***Amps:*** *It is the current that the module generates at STC under 'standard' sun. With the increase of module temperature, the 'amps' drop off; in most cases, one cannot realize 100% of the 'rating' specified on the module,*

(ii) ***Volts:*** *It is the 'working' voltage at which the power output is maximum, though all the modules are made to 12 volts output. It is usually in the range,16.0 to 19.0volts,*

(iii) ***Rated Power:*** *It is the product of 'amps' and 'volts' in the best situation. If a module is rated at 5 amps at 17 volts, then it is 5x17=85 watt module and is expressed in peak-watts (Wp). But in practice, this much of power is not obtained from the module. As an illustration, consider the case of a battery charging at 12 volts.*

If the current from the module is 5.1 amps, the power going into the battery is 5.1x12~61 watts. i.e.,24 watts of power (85-61) is lost; actually this power is not lost,- it is not being produced. The situation will be worse if the battery is low, say 11 watts. This rating is at 'STC' and in general it is never observed.

(iv) ***Short-circuit current:*** *If the module is exposed to sun, the reading in the amp-meter connected across the module denotes short-circuit current. This is needed for sizing the wiring, fuses and breakers. This reading is about 10% higher than the rated 'amps'.*

(v) ***Open-circuit voltage:*** *It is the voltage that the module puts out under zero load. This parameter is important in designing higher voltage systems, because it indicates the maximum possible voltage when the modules are disconnected. It usually varies from 18 to 25volts; a 48 volt system can go as high as 100 volts open-circuit.*

3.3 POWER FROM THE MODULE:

The output power from a module is a very important parameter. The power delivered by a 12 volt module producing 4 amps of current is 48 watts. If two of these modules are connected in series, the voltage gets added with no change in current, delivering a power of 96 watts (=24 × 4);on the other hand, if the modules are connected in parallel, the current gets added while the voltage remains the same, delivering the same amount of power, 96 watts (12 × 8). These two configurations are illustrated in Fig.3.2. The 'rated' power is delivered under STC. But the maximum sun intensity of 1000 watts/sq. m is seldom realised.

The sunlight reaching the module is limited by three factors:(i) weather conditions such as clouds, fog, pollution, (ii) the position of the sun in the sky and (iii) the sunshine hours in the day. To elaborate the first factor, a 50 Wp module should produce 50 watts for each hour of sun shine at 1000 watts/sq.m.

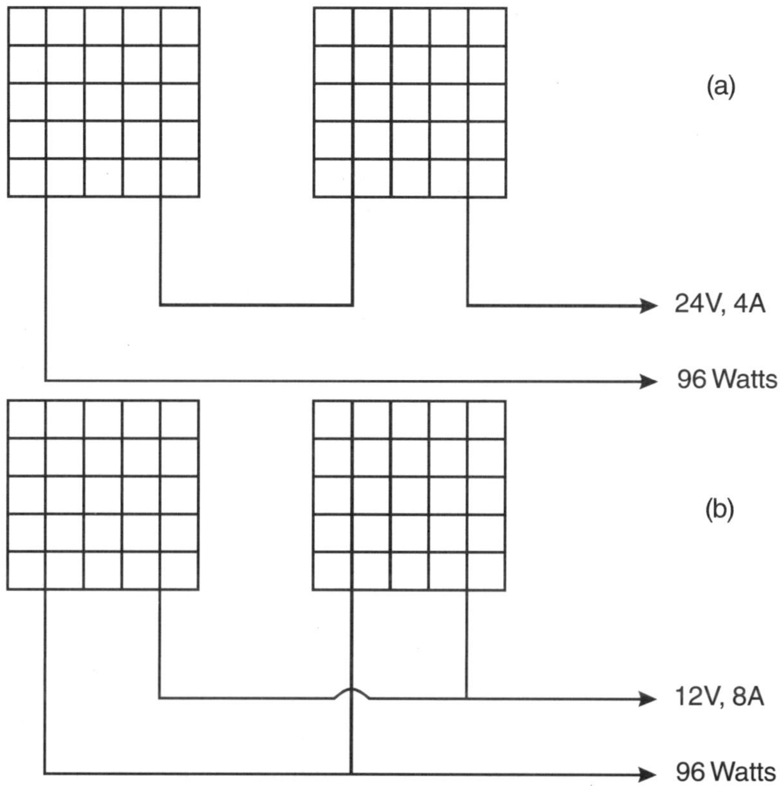

Fig. 3.2 *Power from modules connected in* **(a)** *Series and* **(b)** *Parallel*

If the light intensity gets to half due to the weather, the power produced also becomes half; on a heavy overcast day, the 'diffuse' sun light available on earth's surface may be around 300 to 100 watts/sq.m. The current and the output power may lower due to reduced sun's energy, but the voltage remains the same. Regarding the position of the sun in the sky, the solar light travels faster through the atmosphere if the sun is very high in the sky as in summer season than when it is low in the sky as in winter season. The daily sun shine hours is very critical for areas away from the equator because the number of sun shine hours between the seasons vary.

3.4 FABRICATION OF PHOTOVOLTAIC MODULES

The designing and fabrication of a PV module is an evolved exercise and a careful processing at every step is essential to ensure high reliability and longevity. The number of solar cells are chosen depending on the level of power that the module is expected to deliver. Since the solar cells are thin and fragile, a mechanical support/backing is required. To protect the solar cells and the electrical circuitry from the environmental effects, the cells along with the circuitry are encapsulated and rigidly fixed to a frame. All the electrically active elements are effectively insulated from any conducting contact and from the ground to protect the people from the possible shock.

The materials for the fabrication of modules, encapsulation of solar cells, interconnects, wiring and terminations and structural support, have to be carefully chosen so that the modules withstand the environmental impacts such as diurnal and seasonal temperature variations, rain, snow, humidity, thermal conditions including freezing effects due to wind and storms. The modules should be stable and able to deliver reliable power over its life span of 25 to 30 years without degradation.

There are no established standards fixed for cell or module size and shape but are decided by the voltage, current and power level of the application. Solar cells of circular, square, rectangular and quasi-rectangular shapes of different sizes are available from different manufacturers. Photographs of some typical cells are shown in Fig. 3.3. The modules are also produced in different dimensions and are shown in Figs. 3.4a and 3.4b. These cells and modules are manufactured by TATA BP SOLAR India Limited. The number of cells in a module also differ, but the criterion is that, in a module, the cells are connected for optimum charging performance at low light levels and hot climatic conditions.

In Fig.3.3 solar cells of 125 mm square, 125 mm pseudo square, 125 mm circular and 100 mm circular are shown (Plate.1). Modules made of mono-crystalline cells of different capacities are shown in Plate 2. Modules made of multi-crystalline (polycrystalline) cells are shown in Plate 3. In Plate 4, the Solon Polycrystalline glass-glass laminate with point fixing is shown.

The flat-plate module designs are many and are broadly grouped into three types based on similarities in their design (Ref.1). They are' glass superstrate',' rigid substrate' and 'double glass' as schematically shown in Fig.3.5. The glass superstrate type is the most commonly used module design. The front cover, the encapsulants, the spacer and the back cover are the main material components required for the module fabrication.

Plate 1 Photographs of Silicon Solar Cells: Left to Right:
125mm Square(Poly Si), 125mm pseudo Square (crystalline Si), 125mm dia. Round (c-Si), 100mm dia. Round (c-Si)

Fig. 3.3 *Photographs of typical solar cells.*

Fig. 3.4(a) Photographs of C-Si solar modules.

Fig. 3.4(b) *Photographs of poly silicon modules*

Plate 4 The Solon Polycrystalline glass-glass laminate with point fixing

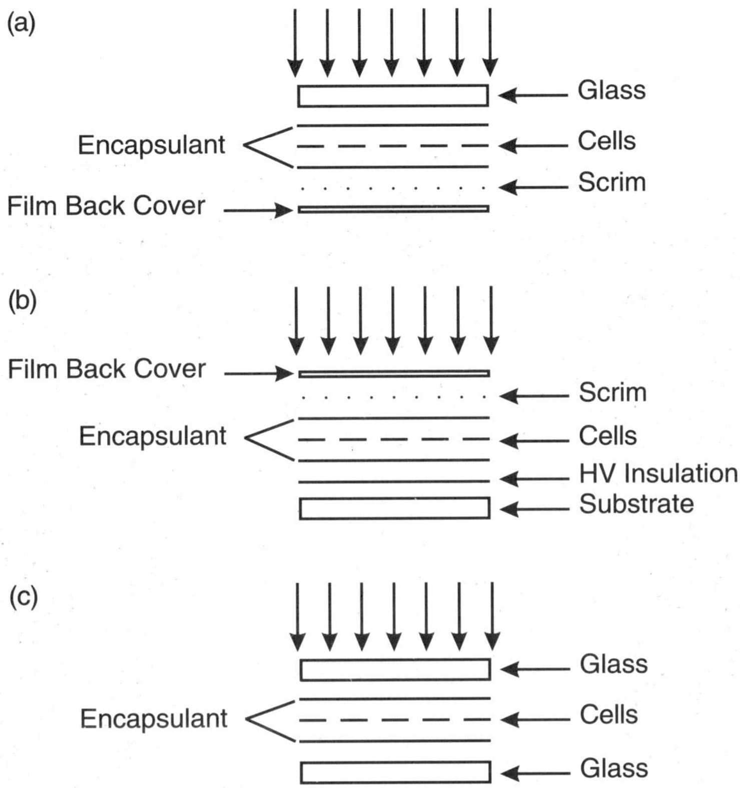

Fig. 3.5 *Flat Plate Module design types:* **(a)** *Glass Superstrate;* **(b)** *Rigid Substrate;* **(c)** *Double Glass*

3.4.1 Module Materials

The materials for the front cover should be

* *transparent in the silicon cell response range, 0.4 to 1.1 micrometer,*
* *abrasion resistant,*
* *cleanable with water*
* *UV screening and oxygen impermeable and*
* *resistant to physical impacts.*

There are four classes of materials, glass, fluorocarbon films, acrylic films and silicone hard coats, which meet these requirements. However, glass is the most favoured front cover material. The front cover functions as the front side weather barrier and front optical surface.

The encapsulant materials are required to provide elastomeric support for the cell circuitry, the optical coupling between the front cover and the cell' surfaces and adhesion to the different layers of materials in the module. These materials should have

* *transparency in the spectral range, 0.4 to 1.1 micrometer,*
* *low elastic modulus,*
* *electrical insulation*
* *resistance to creep over the operating temperatures.*

Further, they should be adherent to other materials used in the module and to adhesives and primers coated on to other materials. Sheet encapsulation, also called lamination is the common practice and the preferred sheet materials are poly vinyl butyral (PVB) and ethylene vinyl acetate (EVA) based formulations.

A porous spacer or scrim is used to create a guaranteed thickness of encapsulant between the cell circuit and the front or the back cover. The encapsulant then provides electrical insulation between them. It further prevents the circuit from touching a film cover which could damage the cover after a number of thermal cycles and also helps to remove air from the module during the lamination. The spacer materials should be

* *porous to allow encapsulant impregnation,*
* *conductive to air flow,*
* *electrically insulating,*
* *relatively incompressible and*
* *transparent in the spectral range, 0.4 to 1.1 micrometer.*

The non-woven glass mats are preferred as spacer materials.

The back cover which primarily functions as backside weather barrier of the module should have

* *high structural strength,*
* *low solar absorption, and*
* *high emissivity in the infra red region in order to reduce module temperature.*

It should also be impermeable to humidity and oxygen to prevent the degradation of encapsulant and corrosion of metals used for cell contacts, and must be abrasion resistant to prevent erosion by particulates in the winds.

3.4.2 Fabrication Process Steps

The fabrication process consists of several steps,- some are performed concurrently and some consecutively. The sequence of processing steps are as follows:

* Inspection of solar cells, visually and dimensionally,
* Evaluation of solar cells by studying I-V characteristics,
* Sorting and matching of solar cells, based on I-V data,
* Preparation of solar cell interconnects and bus bars and cleaning,
* Front side solar cell interconnect tabbing,
* Post tabbing, cleaning of cells and inspection for damages,
* Module interconnection and bussing, and visual inspection of modules,
* Preparation of components for lamination, and cleaning of pre-assembled lamination components,
* Stacking of the laminated components,
* Vacuum lamination of the module and temperature curing,
* Assembly of supporting frames, gasketting & edge sealing, and fixing of terminal box,
* High voltage insulation test,
* Module performance test by studying I-V curves,
* Module inspection, visually & dimensionally, and cleaning of modules,
* Packing and storing.

Details of each of these steps are given in Annexure 3.

3.5. MODULE TESTING

The finished module is tested for both insulation and performance before it is packed. The voltage insulation test is performed to measure the leakage current from the cell circuit to the ground (i.e., frame) of the module using a DC insulation tester. The insulation resistance of the module shall not be less than 100 mega ohms after application of maximum voltage of 1000 volts dc, for a duration of one minute, if the module to be accepted for use. Incidentally, no dielectric breakdown or surface texturing should happen.

The module electrical performance test is done to determine peak power output under standard conditions of irradiance and temperature. This is obtained by studying I-V characteristics using a sun simulator as a source of energy and measuring the incident irradiance with a 'reference' solar cell. (Annexure 4). Solar cells chosen as 'reference' cells shall be highly stable having essentially the same 'spectral response' as the test samples.

The 'reference' cells are calibrated by an approved Solar cell calibration Agency, by measuring the short-circuit current per unit of standard sun light irradiance (Amp/Watt.-sq.m.).

The continuous power output from the modules over a period of time is very important feature of a PV system and is decided by the solar energy input and the efficiency of power conversion. The efficiency is the power output under a given set of conditions that is measurable directly. The peak efficiency, n_p is the most useful parameter that can be measured by determining the output power at peak rating conditions,- 1000 Watts/sq.m at AM 1.5 and cell temperature of 25°C. It is given by

$$\eta_p = \frac{\text{Power Rating}}{1000\,(\text{Watts/Sq. mt})} \times \text{Module Area} \quad \text{.......... (1)}$$

From the peak efficiency, 'operating efficiency'(η_{op}) of the module at the installation site with the existing operating conditions can be deduced (Ref. 2)

$$\eta_{op} = (\eta_p) - \tau(\text{NOCT} - 25^\circ\text{C}) \quad \text{.......... (2)}$$

where NOCT is the 'nominal operation cell temperature' of the module and τ is the coefficient of energy loss per degree celsius.

For crystalline silicon, τ is about 0.4 to 0.5 % per degree celsius. If NOCT is taken as the weighted average temperature at which most of the power is delivered, say 40 to 50°C, n_{op} works out as 9% if n_p is 10%.

3.6 SELECTION OF MODULES

Several manufacturers of PV modules are spread over globally, the leading ones being from Japan, USA and Germany. Other countries such as UK, Australia, Spain, Italy, India, Canada, Tiawan produce PV modules. In recent years China has emerged as a very strong competitor to the top producers of Silicon Solar cells and Modules. In India, the key manufacturers are Bharat Heavy Electricals Ltd, Bharat Electronics Ltd, Siemens India, Central Electronics Ltd., Tata BP Solar using mono-/poly- Si cells.

The thin film Industry has shown significant growth and thin film modules of Cd Te, CIS, a-Si tandum, and μc Si/a-Si, are available in the market. The world maps of the producers of crystilline silicon cells and modules and of thin film modules are published by S & W Energy recently (Ref. 3).

Each manufacturer provides information about the modules which differ from one manufacturer to the other. Some representative data, is furnished by each manufacturer to ensure proper selection by the user. The data provided generally are, Rated power, Open-circuit voltage, Short-circuit current, Maximum power voltage, Maximum power current, Efficiency at max power point at STC, temperatures at which the module to be operated and temperature coefficient of efficiency.

In addition, typical I-V curves of the module are also provided. Fig. 3.6a, b show the I-V curves of modules manufactured by Sun Power corporation, USA and SHARP, Japan as typical examples.

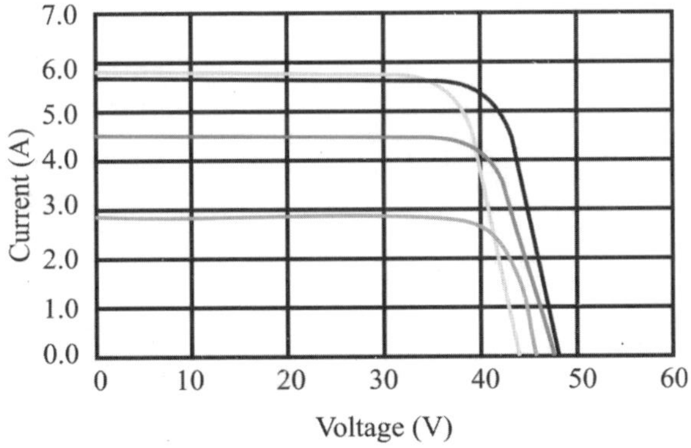

Fig. 3.6(a) *I-V curves of Sun Power Corporation Modules (215 W)*

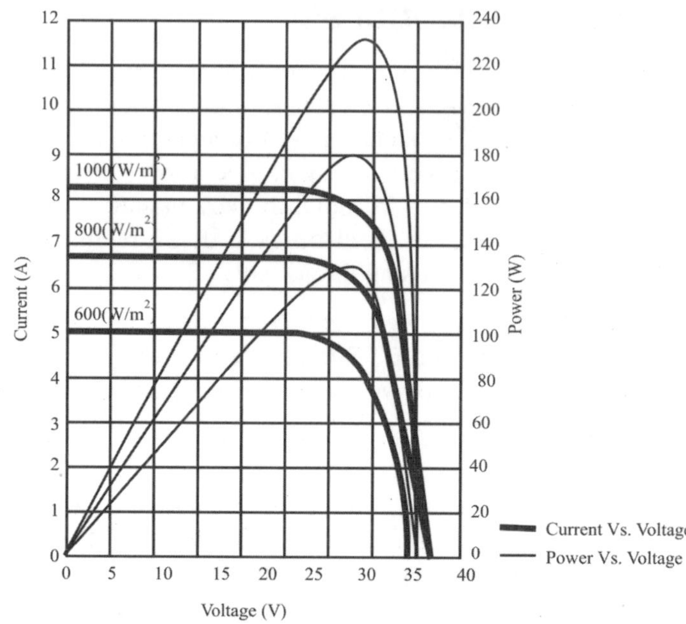

Fig. 3.6(b) *I-V curves of SHARP Modules (230 W)*

Table 3.1 *Sun Power Corporation Modules*

Measured of Standard Test Condition (STC) Introduce of 1000 v/m^3, air mass 1.5 g and cal temperature 25 °C		
Peak Power (+/–5%)	Pmax	215 W
Rated Voltage	Vmp	39.8V
Rated Current	Imp	5.40A
Open Circuit Voltage	Voc	48.3V
Short Circuit Current	Isc	5.80A
Maximum System Voltage	IEC, UL	1000V, 600V
Temperature Coefficients		
	Power	–0.38%/°C
	Voltage (Voc)	–136.8 mV/°C
	Current (Isc)	3.5 mA/°C
Series Fuse Rating		15A
Peak Power per Unit Area		173W/m^2, 16.1W/m^2
CEC PTC Rating		198.5W

Table 3.2 *Sharp Modules*

Maximum Power (Pmax)*	230 W
Tolerance of Pmax	+10%/–5%
Type of Cell	Polycrystalline silicon
Cell Configuration	60 in series
Open Circuit Voltage (Voc)	37.0 V
Maximum Power at Voltage (Vpm)	30.3 V
Short Circuit Current (Isc)	8.24 A
Maximum Power at Current (Ipm)	7.60 A
Module Efficiency (%)	14.1%
Maximum System (DC) Voltage	600 V
Series Fuse Rating	15 A
NOCT	47.5 °C
Temeperature Coefficient (Pmax)	–0.485% / °C
Temeperature Coefficient (Voc)	–0.36% / °C
Temeperature Coefficient (Isc)	0.053% / °C

*Measured at (STC) Stadard Test Conditions: 25 °C, $1kW/m^2$ insolation, AM 1.5

In selecting a module, it is also desirable (i) to examine the best price per Amp than the best price per watt and (ii) to look into the size of the modules because low efficiency modules are bigger in physical dimensions and require bigger physical structure and larger area for installation. Most manufacturers of PV modules are fabricating to exact international and/or industry standards that ensures a life span of at least 25 to 30 years.

3.7 GLASSLESS PV MODULES

New materials for encapsulation, i.e., front and back covers have been recently studied for use in 'glassless' PV modules for use in new applications specially in building integration. The general features of these modules are the same as for the standard modules, namely, flexibility, weatherability, electric insulation and resistant to environmental effects.

The new materials developed for back cover (Ref.4) are phenolic hard paper (a rigid laminated sheet, 8 mm thick) and PVF/PET/PVF flexible laminated film of thickness,0.27 to 0.44 mm. The front cover materials apart from being highly transparent shall have high optical transmission, temperature and mechanical stability, good adhesion to EVA, long term stability and good water vapour permeability. SiO_x film can absorb UV radiation depending on Si/O ratio, and has a good optical behaviour.

It also prevents rapid aging of solar cell arrays. Several substrate films have been studied on their suitability for barrier coatings of SiO_x to obtain water vapour permeability < 0.1 g/m^2d because water vapour affects the efficiency and life of the modules. The substrate material films found to be suitable for depositing SiO_x layers are ETFE, PET, PEN and PET/PEN in that order.

The heat shrinkage measurements, adhesion to EVA and light transmission of these materials are thoroughly investigated. ETFE films are susceptible to wrinkling due to temperature; the lamination process is suggested to be done as low temperature process to avoid wrinkling. The new materials are of high quality and can enhance the reliability of the modules. Cost-wise too, they will be cheaper when compared to materials presently in use.

REFERENCES

1. ***R.G.Little et al**: Solar Cell Module Construction and Mass Production Process, 1st Intl. PV Science & Engineering Conference, Kobe, Japan (1984)*
2. ***The Design of Residential PV Systems**: Vol.8, Sandia National Labs, Albuquerque, NM, USA (1988)*
3. ***Sun & Wind Energy:** PV Industry:Growth across the board,p118(4/2008)*
4. ***Sabine Degiampietro et al**: Proc. 16th European PV Solar Energy Conf. p2327 (2000)*

CHAPTER FOUR

SOLAR PHOTOVOLTAIC SYSTEMS

4

SOLAR PHOTOVOLTAIC SYSTEMS

4.1 INTRODUCTION

If more power than can be provided by a single photovoltaic module is required for any application, then a photovoltaic system is designed by linking more modules normally operating at 12 volts dc. The number of modules are decided by the power requirement. A photovoltaic system broadly consists of (Fig. 4.1):

* Photovoltaic Array, comprising a few numbers to a few hundreds of modules,
* Charge controller, to regulate the power from the PV array,
* Power storage system, consisting of deep cycle batteries,
* Inverter, to convert the dc power from the array into ac power,
* Cables, sensors, physical structure, and
* Backup power supply or linking to utility grid, if needed.

The daily energy output from PV array will vary depending on its size, orientation and location, season of the year and the daily weather conditions. On average, a module produces about five times its related power output in watt-hrs/day in summer and about two times that amount in winter.

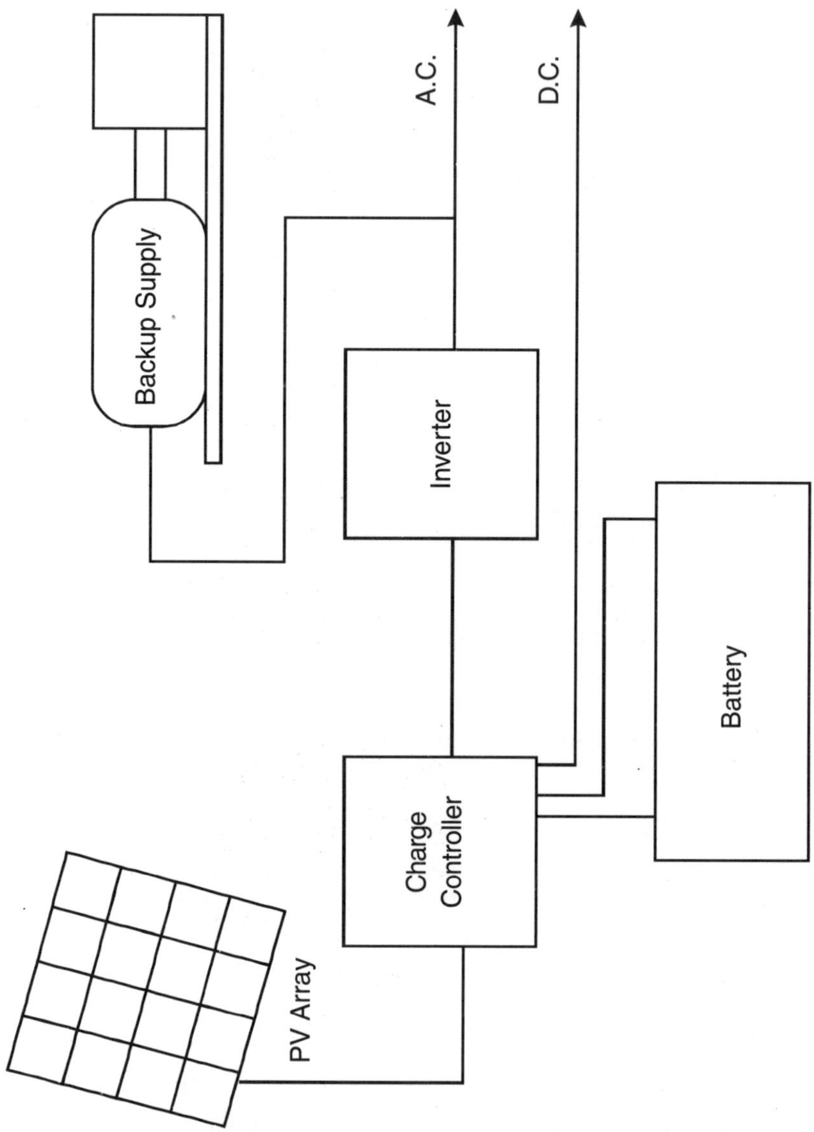

Fig. 4.1 *Elements of PV System*

Energy storage is necessary because the power may be required when sun is not shining. Specially designed 'deep-cycle' batteries, lead-acid type, are generally employed for storage systems. Battery sizing depends on the application, the daily solar radiation, the total load, peak load and the number of days for which storage required.

The charge controller is placed between the PV array and the storage battery to prevent the battery from being damaged either due to overcharging or over-discharging and to enhance its life.

The power from the PV system is dc and most of electrical appliances work on ac. The device, "inverter" is used to convert 12 volts or 24 volts dc power into 220 volts or 110 volts ac power.

There will be some loss of power in this conversion because no inverter is 100% efficient. A backup or auxiliary power supply or interface with utility grid may also be necessary, if uninterrupted power supply is required. The batteries, charge controllers and inverters are dealt in detail in the following chapters.

4.2 TYPES OF SOLAR PV SYSTEMS

Solar PV systems are broadly classified into three types, each type is characterized by unique considerations in design, installation and operation.

4.2.1 Stand-alone Systems

This type of system comprises an array of PV modules, Fig. 4.2, connected to a battery bank through a charge controller. The charge controller switches off the PV array when the battery is fully charged and switches off the load when the battery becomes discharged. The battery bank capacity must be large enough to store the charge produced during the day for use in the night. In isolated places which are off the grid, the battery bank stores energy produced during sunny days for use during cloudy and rainy days. There is no backup power supply in this system nor connected to the utility grid.

The system works on only PV technology. It is highly applicable to remote places not connected by grid where domestic power requirements or of other applications are not large.

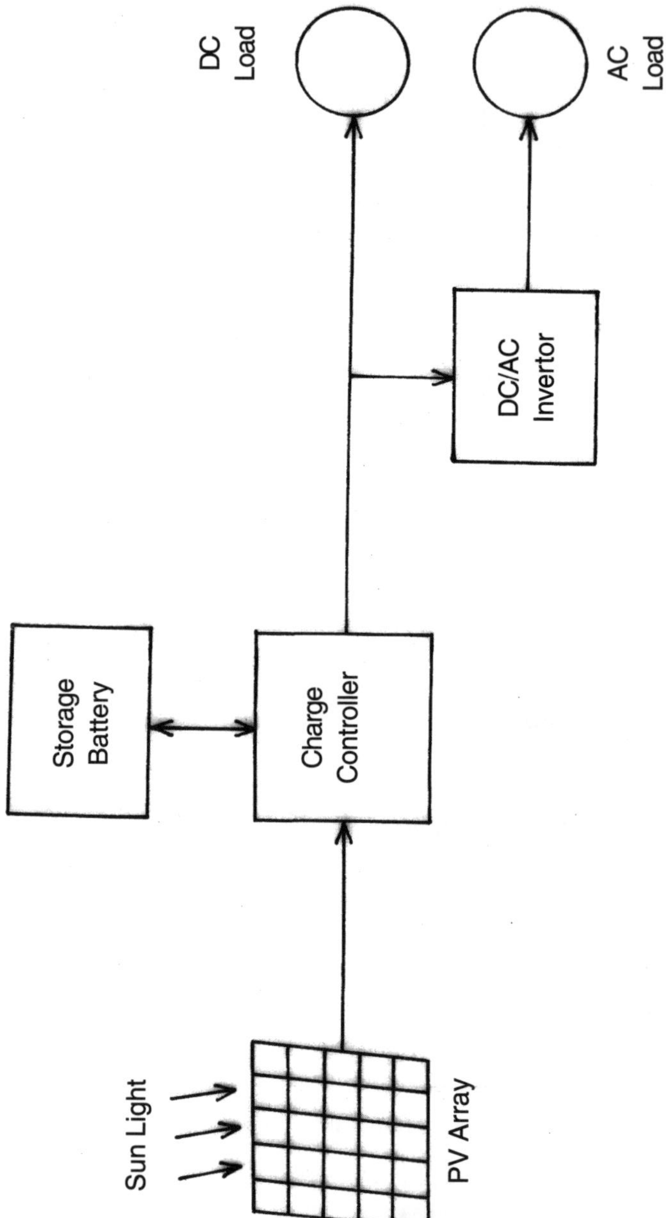

Fig. 4.2 *Stand-alone PV System*

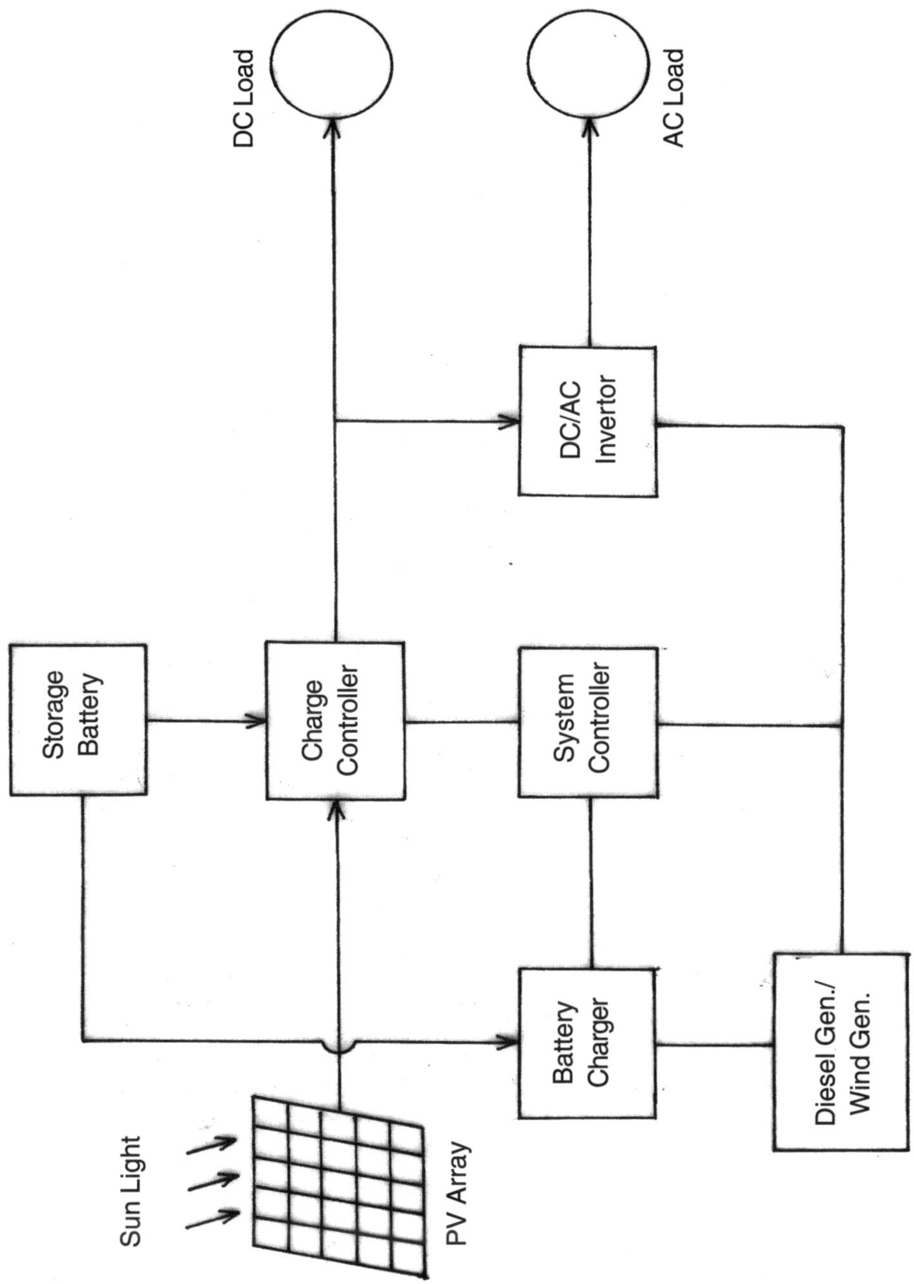
DC Load
AC Load
DC/AC
Invertor
Storage
Battery
Charge
Controller
System
Controller
Battery
Charger
Diesel Gen./
Wind Gen.
Sun Light
PV Array

4.2.2 Hybrid PV Systems

This system consists of a combination of a PV array and another form of energy source such as diesel generator, small hydro plant or wind generator or any other source of electrical energy(Fig. 4.3). This system requires more sophisticated controls compared to stand-alone systems. The PV array and the battery bank are sized to supply average energy needs throughout the year, and the auxiliary supply is used to supplement during low output periods and also in periods of prolonged bad weather. The cost of the system is generally less compared to stand-alone system. The hybrid systems are highly suitable and can offer a reliable source of AC power for rural areas and other remote applications.

4.2.3 Grid-connected PV Systems

The PV array is connected to utility interactive type inverter and utility power grid (Fig. 4.4). The array produces dc which is converted into ac by the inverter. The power is consumed by the load and/or pumped into the utility grid network. Batteries are not included in the system. These systems are built as small power stations to pump the power produced into the grid. They are also installed in buildings which have utility grid facility, so that when solar power is not adequately available, they can switch over to the grid power. Since the power is not generated in the night, the load operates from utility power.

The inverter shuts down automatically as a safety measure at the time of grid power failure, and reconnects automatically when grid power resumes. Grid connected PV systems are preferred if PV power is to become part of daily life. This type is highly suited for building integrated photovoltaics.

The advantages of the grid connected systems are: (i) no energy storage facility and consequently no losses due to the battery storage, (ii) no energy wastage because surplus electric power is pumped into the grid, and (iii) no sizing problems because one can start with a small system and as the demand grows, the capacity can be increased. All problems regarding charge controllers, batteries, diodes and sizing become insignificant in this system. This design is highly suited to residential and to smallsize commercial buildings and also for places where power outage is frequent.

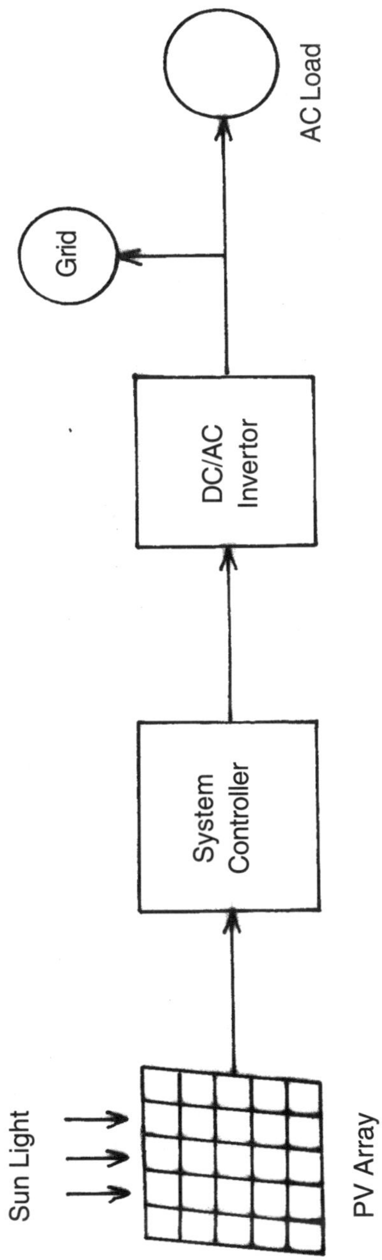

Fig. 4.4 *Grid-connected PV System*

Of the three types of PV systems, stand-alone PV systems are the most favoured for remote places which are off the grid; and hybrid systems are desirable when the energy costs have to be minimal.

4.3 PV ARRAY TESTING FOR FAULTS

Small PV array is a simple electrical system, but a large one can be complex with many connectors and controls. It is difficult to know when the system is working below its full capacity and may take months before the problem becomes clear. A large array of PV modules consist of many strings with each string containing many modules connected in series and a diode. Each module consists of many cells connected in series and sometimes in parallel also, and a bypass diode. These components have interconnections including sub-array strings, junction boxes, different type of cables etc. Faults can therefore arise at any part of the array. The measurement of 'open-circuit voltage' and 'short-circuit current' may identify certain faults only, but the I-V characteristic may identify all the complex interactions and faults. An 'Array Tester' is used for measuring I-V curve and for trouble shooting; it can narrow down the faulty locations. It is desirable and also effective to measure the entire system at one particular point than to measure each part separately using meters.

The array I-V curve given in Fig. 4.5(a) shows the 'expected' and the 'actual' values of power output from the array (Ref.2). The actual power output is 20% less than the expected output. This fault could not have been noticed till the cumulative output was found to be lower than the specified value probably after a long time of operation (say, annual check), if this method of measurement was not performed.

Errors in the wiring during installation are the common faults that occur in the first few weeks of operation because the weathering and thermal fluctuations can expose these poor connections. Fig. 4.5(b) reveals the presence of faults due to wiring errors in commissioning. These affect the efficiency of the system in the initial period of installation that is not otherwise noticed till the normal annual check is done. This figure indicates a substantial decrease in the 'expected' output power due to the late discovery of a problem that occured early.

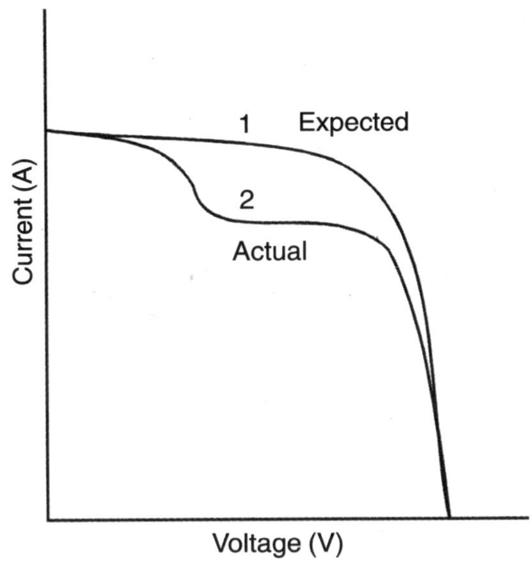

Fig. 4.5(a) *I-V Curve of an Array showing "actual" and "expected" power outputs*

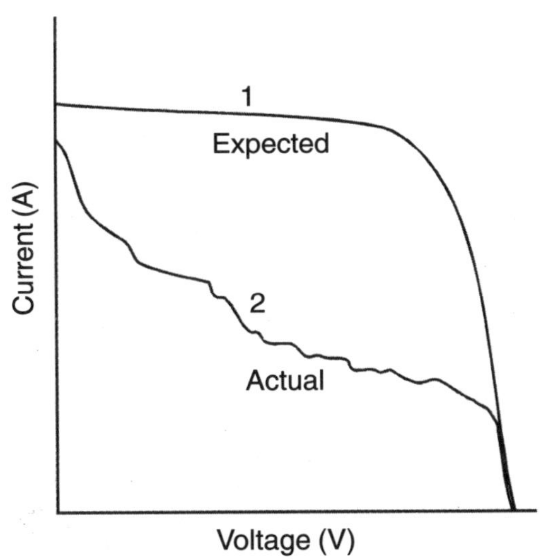

Fig. 4.5(b) *I-V Curve of an Array showing the presence of "commissioning" faults*

Measurement of the open-circuit voltage and short-circuit current using standard meters do not identify 'power related' faults or problems because they are measured at zero power conditions.

4.4 SYSTEM PERFORMANCE

An approximate estimate of the system output can be made from the daily solar irradiances (kWh/sq.m) and the nominal array rating (kWp) using the expression,

System energy output/day = (PR)(Nominal array rating) (Solar irradiance/day) .... (1)

where PR is the 'Performance Ratio'

which depends on the system design and the application.

The PR can be explained as follows: The output of the PV systems of different sizes (capacities) utilized in different applications may be compared by normalizing them with respect to system size and comparing the monthly mean values of their daily final 'yields'.

i.e., Final yield = Energy supplied per kWp of installed array power (kWh/day / kWp) .... (2)

Combining equations (1) and (2),

PR = final yield/solar irradiance .... (3)

Grid connected PV systems achieve highest performance ratios and 'yields' because all the electric power they produce are either consumed by the load or pumped into the grid. If a high efficiency inverter is used and the array losses are kept to minimum, PRs of about 80% may be produced.

Hybrid PV systems produce slightly lower performance ratios because storage battery and control losses get added to array losses in this type. The PRs are generally around 50 to 70%.

Stand-alone PV systems achieve the lowest performance ratios of about 20 to 30% because their yields are controlled by the sizes of their loads. These systems are sized to meet the full energy demand in winter, though they are oversized in summer.

4.5 FEATURES SPECIFIC TO SPV POWER SYSTEMS

The Solar PV power generation systems are characterized by certain features in contrast to other methods of electric power generation.

(a) *SPV array produces 'dc power'*

(b) *The module/array is neither a constant-voltage source nor a constant-current source,*

(c) *The performance and power output from the array depends on the uniformity of solar illumination on the surface of the cells, the weather conditions, the angle at which the solar radiation is incident on the array and the connected load,*

(d) *The power output from the system is influenced by, whether the PV array is fixed or made to track the relative motion of the sun,*

(e) *The average daily performance depends on the optimization of the tilt angle with respect to the horizontal and its azimuth angle, for a fixed tilt array*

(f) *The array output power differs from the sum of the outputs from individual modules due to mismatch and other losses,*

(g) *SPV modules are modular in nature which provides for the alteration of array capacity at any time, even after installation,*

(h) *SPV systems have merits of being silent, non-polluting and having no detectable emissions,*

(i) *SPV systems are less vulnerable to severe weather conditions.*

These features make the solar PV systems highly site-specific and load-specific. These power systems are preferred to keep the environment clean and to sustain eco-balance.

The sizing and design aspects of the systems are dealt in later pages.

REFERENCES

1. ***Photovoltaic Technologies and Their Future Potential:*** *Produced by Energie-Analagen Berlin GmbH, Berlin, for The CEC, Director General VII for Energy (1993)*
2. ***Bruce Cross:*** *Renewable Energy World, Vol.3, No.3, p111 (2000)*

CHAPTER FIVE

STORAGE BATTERIES

5

STORAGE BATTERIES

5.1 INTRODUCTION

Storage battery is an important and most expensive component in a solar PV system, and it stores electrical energy generated by the PV array. The deep cycle batteries are preferred for PV applications because of their long life and other advantages compared to other types of batteries.

Lead-acid batteries are commonly used in deep-cycle applications; the lead acid battery has developed into a versatile and extremely reliable energy storage device over the last century. Within the catagory of lead acid, lead antimony and lead calcium are the most common. The major types of the lead acid batteries are (i) standard flooded (wet), (ii) Gelled and (iii) Absorptive Glass Mat (AGM) types. The Gelled and AGM types are two distinct concepts in the battery design to immobilise the electrolyte within the cells of the battery, using 'gel' electrolyte in one case and AGM separators in the other.

The 'flooded' batteries may be standard with removable caps and are 'maintenance-free'. The 'gelled' batteries are either 'sealed' or 'valve-regulated' whereas all AGM batteries are 'sealed' and 'valve-regulated'.

5.2 BATTERY LIFE AND DOD

A battery cycle is generally considered to be discharging from100% of charge to 20% and then charging back to 100% of charge. The life of the battery depends on the Depth of Discharge (DOD) cycle and the most common DODs are 10%, 20% and 50%. For instance, a battery may have 20-year life at 5% DOD. But the same battery, if cycled deeper regularly, the life may decrease to 5 years at 50% DOD. Battery life is directly dependent on how deep the battery is cycled each time. If a battery is discharged to 50% every day, it will last about twice as long as if it is cycled to 80% DOD. As a standard, the practical rating suggested is 50% DOD on a regular basis though higher ratings of DOD are not prohibited once in a while. It means that for the estimation of the battery capacity or size for a solar PV system, the DOD is generally taken as 50% for the best storage Vs cost-factor.

5.3 TYPES OF BATTERIES

All rechargeable batteries, in common use, are lead-acid type. The acid is typically 30% sulphuric acid and 70% water at full charge. Whenever power is required over a longer period of time, the battery is designed to be 'deep-cycled'; ie, it is designed to be discharged down to 80%, time after time. Deep-cycle battery, compared to others, has much thicker plates; thicker plates are directly related to longer life,- other factors being equal while the battery with thickest plates has longest life.

i. Sealed batteries are made with vents that cannot be removed. The maintenance-free batteries are also 'sealed', but are not usually leak-proof. Smaller deep-cycle batteries use lead-calcium plates for increased lifespan, while most industrial batteries use lead-antimony for greater plate strength. The lead-antimony type has a very long lifespan, so also higher self-discharge.

ii. *I*n Gelled-batteries, silicon gel is added to turn the acid into a solid mass giving an appearance of 'gel'. The main advantage of these batteries is that the acid is not spilled even when they are broken. The disadvantage is that these batteries have to be charged at a slower rate (C/20) to prevent damage of the cells. Also, these batteries must be charged at a lower voltage than 'flooded' or 'AGM' batteries. If overcharged, voids can develop in the gel affecting the capacity of the battery.

iii. The new type of sealed battery uses 'Absorptive Glass Mats (AGM)'between the plates. These plates possess all the advantages of 'gelled' batteries, but with none of their demerits. AGM batteries have several other advantages over both the 'gelled' and flooded' types. Since all the acid is contained in the glass mats, they cannot spill even if they are broken. These batteries could withstand shock and vibration because they are so designed that the plates are tightly packed and rigidly mounted. They are not susceptible to damage from freezing since there is no liquid to freeze, or to expand due to fluctuations in temperature. The AGM batteries are 'recombinant' and the recombining is more than 99% efficient, ensuring no loss of water. The charging voltages remain the same as for any standard battery. The internal resistance is low enough to avoid heating of the battery even under heavy charge and discharge currents. These batteries, further, have a very low self-discharge, around 1 to 3% per month. The merits of these batteries are: maintenance-free, completely sealed against fumes, hydrogen or leakage and no-spillage even if they are broken. However, cost-wise, they are about 2 to 3 times expensive than the 'flooded' batteries of the same capacity.

5.4 BATTERY CHARGING

The battery capacity is referred in terms of 'ampere-hours' it can hold. An amp-hour (AH) is equivalent to carrying one ampere current for one hour or 10 amps for one-tenth of an hour. This 'amp hour capasity indicates the energy storage capacity of the battery which is the basis for "rating" storage batteries. To keep the battery safe, an optimum charging rate is desirable. The factors that limit the rate of charging are the excessive temperature rise and excessive gassing of hydrogen and oxygen. Generally, the rate for recharging a battery is equal to the amp-hours remaining to put back into the battery; ie, if a 200 AH battery is fully charged, it will take a 200 amp charge-rate, and if 50% is discharged, a 100 ampere charge-rate.

Ordinarily, batteries can be charged quickly and easily up to about 80% charge, and the problems arise only when charging the last 20 %. Most of the battery chargers are provided with a tapering current charge to overcome this problem. In the case of PV, the charge controller performs this function by limiting the current flowing into the battery as it approaches full charge.

Consistent undercharging is also not healthy to the battery because it causes 'sulfation' (formation of lead sulphate layer on the plates) which restricts the battery life. This can be avoided by applying an equalizing charge, about 10% higher than the normal full charge voltage for about 3 to 16 hours, which prevents undercharging of individual cells and ensures that all the cells in the battery are equally charged. During this time the voltage and specific gravity should remain unchanged. The equalizing charge is applied to batteries which are daily deep-cycled, at 10 to 40 days. The 'gelled' and 'AGM' batteries, however, need to be equalized 2 to 4 times only in a year.

Batteries are charged in different ways in different methods; for instance, AC battery chargers, wind chargers, solar PV chargers, hydro-chargers.

5.5 STATE OF CHARGE

All lead-acid batteries supply 12.6 to 12.8 volts for 12 volts when fully charged. If they are stored for longer periods, they loose charge through self-discharge (or leakage) ranging from 1 to 15% per month, depending on the type and age of the battery and the temperature. This leakage would be more if the battery is stored in a partly discharged state for a few months.

During charging and discharging at higher currents, the losses are higher if the internal resistance of the battery is higher; ie, if the battery is discharged faster, the AH capacity is lower and if it is discharged slower, the AH capacity is higher.

The reverse electro chemical process by which a lead acid battry stores electrical energy is represented by the equation (Ref. 1),

$$PbO_2 + 2H_2SO_4 + Pb \underset{\text{discharging}}{\overset{\text{charging}}{\rightleftarrows}} 2PbSO_4 + 2H_2O$$

The state of charge for a typical 12 volt battery at 25°C is given in the following table(Ref. 2). The voltages given are for batteries that are at rest for at least three hours. The voltage measurement is only an approximate indication whereas specific gravity measurement is a better means, though its measurement is not possible for many batteries.

State of charge(%)	12V battery	Volts/cell
100	12.70	2.12
90	12.50	2.08
80	12.42	2.07
70	12.32	2.05
60	12.20	2.03
50	12.06	2.01
40	11.90	1.98
30	11.75	1.96
20	11.58	1.93
10	11.31	1.89
0	10.50	1.75

Source: http//www.windsun.com/Battery/Battery-FAQ.htm)

5.6 EFFECT OF TEMPERATURE

The battery capacity increases with rise in temperature and reduces as the temperature falls. The standard rating is specified at 77 F(25 C) where the battery performs ideally. At freezing, the capacity is reduced by 20% and at about -27 C, the capacity drops to 50%. At +50 C, the AH capacity would be higher by about 12%.

The battery charging voltage is also affected by the temperature; for a 12 volts battery, it varies from 16.4 volts at -40 C to 13.8 volts at +50°C. This variation in battery charging voltage demands for temperature-compensation on the charge controller, if the battery has to operate over a wide range of temperatures. Some charge controllers have built-in temperature-compensation; however, they work well provided the controllers are subjected to the same temperature variations as the batteries.

The battery life gets shortened with increase in temperature. For every 15 F(or 8°C) approximately, over a 25°C cell temperature, the battery life is reduced by half at very high temperatures. The lifespan, at the same temperature, is increased if temperature drops below 25°C, though the capacity is reduced. This is true of all three types,- sealed, gelled and AGM types of lead-acid batteries. In places of extreme heat and cold, the batteries may not use acid of same strength,- higher strength for cold climate and weaker strength in hot climate are recommended.

5.7 BATTERY FOR PHOTOVOLTAIC APPLICATION

The battery as a storage device of electric charge is very critical to PV systems. The manufacturers of all types of batteries used in PV systems offer a number of desirable performance attributes like, deep cycle, high power, heavy duty, vibration-free, maintenance-free and extended life. In choosing a battery for different applications, the specifications given by the manufacturer have to be carefully examined.

Generally, the specifications are:

* operating temperature range (-15 to +50°C)
* self discharge rate
* cycle life to 80% DOD

* charge efficiency from 20% discharge condition
* AH capacity at 10 hr, 20 hr and 100 hr rates(C10,C20 and C100)
* required frequency for topping up the electrolyte
* resistance to overcharging
* physical dimensions and weight

The 'flooded' type are commonly used in PV systems. It has a high acid volume, about 27% wt, and the positive plates are protected to ensure fairly high reliability in cyclic applications. Lead-calcium-tin alloys are used in the grids for the plates which eliminates the need to add water making the design maintenance-free.

The most attractive lead-acid battery for use in PV systems is 'flooded type' made with tubular plates and grids with low Sb-Se content. These batteries are characterized by

- *low self-dischrge rates, about 20% per month at 25°C.*
- *appreciable life-cycle, around 1800 cycles at 80% DOD at 40°C, and*
- *wide operating temperature range, -15 to +50°C.*

Good quality batteries of this type work for a minimum of 5 to a maximum of 8 years if properly maintained. These batteries are also moderately priced. However, a particular disadvantage with these batteries is that they do not accept low rates of charging.

AGM valve-regulated type are very sensitive to operating temperatures and do not provide reasonable service life at higher temperatures. However, 'gel' type are found to be eminently suited for PV, although the shortage of electrolyte in this type, a characteristic feature of the technology, results in the much lower battery capacity compared to 'flooded' type. The valve-regulated lead acid batteries use lead-calcium-tin alloys for the grids and are maintenance-free.

Ni-Cd batteries are also suggested for PV applications, but they are highly expensive, and of low efficiency (65 to 80%). Their cycle life is no better than the lead acid battries, which makes them not viable for systems with a daily cycle.

Ni-Fe batteries also have as many drawbacks as advantages when compared to lead-acid batteries. They are low efficient, typically 60-65%, sometimes as low as 50%, very high rate of self discharge, high gassign/ water consumption, high internal resistance, and can reduce the overall efficiency of the solar system as much as 25%. The output voltage varies with load and charge much more than other batteries. If an invertor is used, it must be capable of with standing to the voltage variations.

Both Ni-Cd and Ni-Fe are considered poor choice for solar applications inspite of their long life and thousands of cycles.

The design features to be considered important for selecting batteries for PV systems are the following:

(a) Thick positive plates(> 2.5 mm) generally give longer life and better deep cycle performance compared to thin plates,

(b) A greater amount of electrolyte is desirable because the amount of electrolyte influences the battery performance at slow rates of discharge, though this is not possible in valve-regulated type,

(c) Positive plates made from lead-calcium-tin alloys require a high positive-to-negative active material ratio to protect them from the effects of deep discharge cycling,

(d) lead-calcium alloys will result in much less maintenance compared to other types.

The selection of a battery for PV applications is generally a trade-off between its desired energy storage capacity, its life, and its cost.

5.8 BATTERY AGING

The battery requirements alter as the batteries get old. Usually, the batteries require watering more often. With aging, the charging time becomes longer and the charging rate higher.

5.9 IMPORTANT GUIDELINES

The storage battery in PV systems requires proper care from the day it is included in the system to the last day of its expected life for providing good service. The users have to be cautious and follow the guidelines listed below:

(a) *Almost all batteries do not reach full capacity until cycled 10 to 30 times; A brand new battery will have a capacity, 5 to 10% less than the rated capacity,*

(b) *Batteries should be watered after charging to cover the plates. After full charge, the water level should be even in all cells, usually 1/4th to 1/2 inch, below the bottom of the fill well in the cell,*

(c) *In a battery bank, the replacement batteries should be of the same size, type and preferably from the same manufacturer. A new battery should not go into a pack containing batteries more than 3 months old and performed more than 75 cycles. Age and usage level should be the same as other batteries in the bank,*

(d) *In the case of 'flooded' batteries, the vent caps should be in their place output of about 1/300th to 1/1000th of the AH capacity has to be chosen,*

(h) *Lead-acid batteries do not have a 'memory' and there is no need to discharge totally to avoid this memory,*

(i) *A new 'flooded battery' should be fully charged and equalized;- the hydro meter reading should be noted for future reference only, and*

(g) *A fully-charged battery of 12 volt capacity, generally shows between 12.5 and 14.0 volts. A battery shall no longer be used when the voltage drops below 11.5 volts; if used, its life gets reduced. For good performance the voltage must remain within this range, 12.5 to 14.0Volts.*

REFERENCES

1. ***Terrance D Paul :*** *How to design an Indepandant Power System, Published by Best Energy Systems for Tomorrow Inc., Necedah, Wisconsin, USA (1981)*

2. *www. windsun.com/battery/battery-faq.htm.*

CHAPTER SIX

POWER CONDITIONING EQUIPMENT

6

POWER CONDITIONING EQUIPMENT

'Power conditioning equipment' by convention, in the context of solar photovoltaics, refers to all the sub systems that are employed to process the electrical energy generated by the SPV array in order to make it directly usable by the customer. Main subsystems in the power conditioning equipment are Charge Controller and Inverter. However, other than these two subsystems, instrumentation plays an important role in the design, construction, safety and maintenance of a PV system.

6.1 CHARGE CONTROLLER

The charge controller in a solar photovoltaic system efficiently and safely interfaces the PV array, battery bank and the load. The main functions of a charge controller are:

(i) *control the charge rate based on the state of charge of the battery,*

(ii) *disconnect the battery from the PV array on battery reaching a preset voltage, to prevent overcharging,*

(iii) *disconnect battery from the load on battery voltage falling down to a preset value to prevent the battery*

being discharged below the maximum depth-of-discharge level,

(iv) *disconnect the load from the system on sensing any abnormal load conditions,*

(v) *set the operating point to enable maximum power transfer from the PV array to the load to ensure maximum energy utilization, and*

(vi) *provide necessary operational status indications in the form of audio/visual alarms.*

Charge controller is essentially a step-down dc-dc Converter that operates on the output voltage of the PV array, and generates dc voltage output that is compatible with the inverter input voltage requirements or direct DC loads.

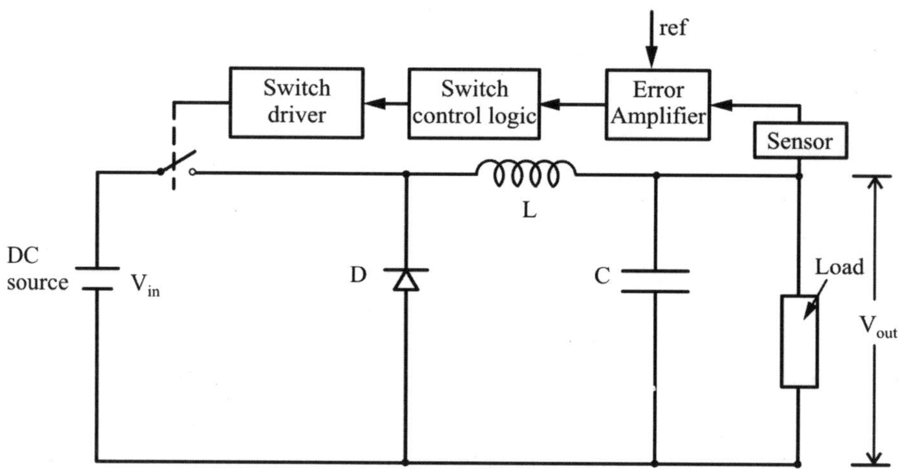

Fig. 6.1 *Basic schematic of a typical step-down DC-DC converter*

Shown in Fig.6.1 is the schematic of step-down dc-dc converter operating on Pulse Width Modulation (PWM) technique. Output voltage is sensed and compared with the desired voltage. The error is amplified and switch control logic sets the PWM signal parameters and the switch is driven on-off accordingly. The switching wave form is shown in Fig.6.2

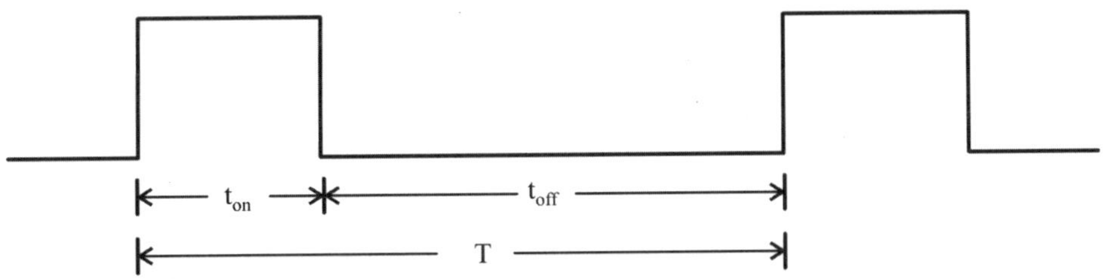

Fig. 6.2 *Switching wave form*

Assuming constant frequency operation,

$$f = \frac{1}{T} = \frac{1}{t_{on} + t_{off}}$$

During the 'on' period of the switch, a voltage, $V_{in} - V_{out}$ appears across the inductor. Assuming all ideal components, the current through the inductor is $(V_{in} - V_{out}) / L$. During the 'off' period, as the current in the inductor cannot change instantaneously, current continues to flow through the inductor into the load, while diode provides the return current path.

Current decreases with a slope, $- V_{out} / L$. The wave forms are shown in Fig.6.3.

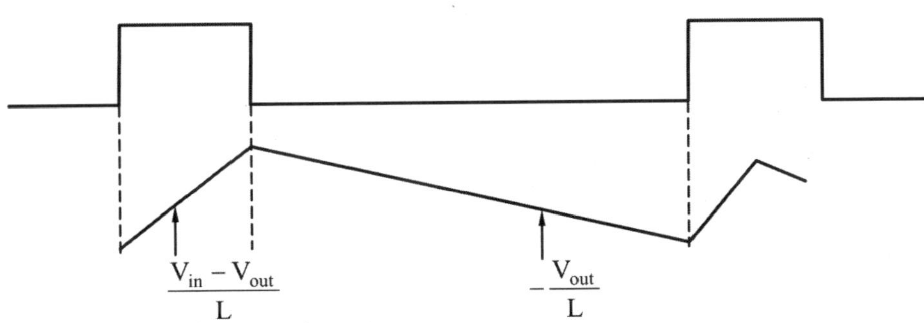

Fig. 6.3 *Current wave forms.*

From an observation of the steady state current wave-form, it can be seen

$$\{V_{in} - V_{out} / L\}. t_{on} = (V_{out} / L). t_{off}$$

Solving for V_{out},

$$V_{out} = V_{in}.\{t_{on} / t_{on} + t_{off}\} = V_{in}. t_{on} / T.$$

t_{on} / T, the ratio of 'on' time to total switching cycle time is called 'duty ratio' (also known as 'duty cycle'), and is generally denoted by 'd'.

$$V_{out} = V_{in}. d$$

V_{out} can be changed by changing the duty ratio. This can be achieved by changing the 'on' time while keeping the frequency (1/T) constant or vice versa.

Performance of Solar PV Charge Controllers:

With the availability of a wide range of dedicated integrated circuits for implementation of PWM and switching devices with very low 'on' resistances and improved switching characteristics, charge controllers employed in Solar PV systems have reached very high efficiencies, in excess of 97%. Charge controllers with wide range of operating voltages and power ratings are commercially available with robust designs suitable for use in PV systems (off-grid, grid-tied and hybrid types) for varied locations.

For maximum utilization of the energy available from the solar generator, it is essential to match the load to the solar generator. It can be seen in Fig. 6.4, that the maximum power point at different solar insolations varies in such a way that the optimum voltage of operation keeps decreasing with reducing insolation. Hence, a Maximum Power Point Tracking (MPPT) charge controller which automatically permits the PV array to operate at the voltage that produces maximum power output is utilized.

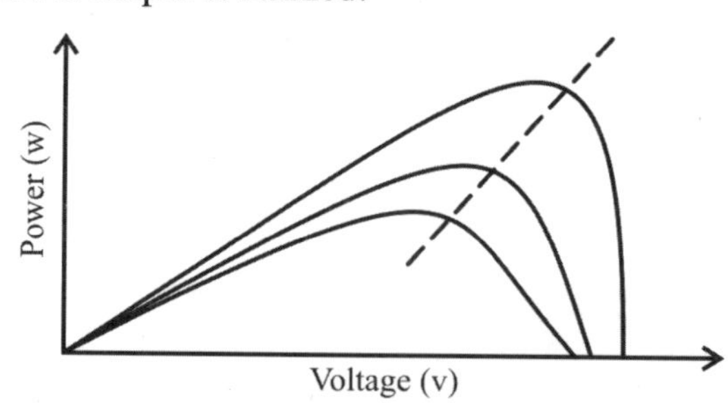

Fig. 6.4 *MPP Tracking at different solar insolations.*

In a typical MPPT charge controller, a high frequency DC-DC converter is used to provide a constant voltage to the battery for most efficient charging, while the input voltage to the DC-DC converter keeps varying with the maximum power point. A schematic block diagram of a MPPT is shown in Fig. 6.5.

Maximum power point tracking can be performed using a pulse width modulated DC-DC converter. By controlling the duty ratio, one can control the load voltage for a given PV array voltage. When the transistor is 'ON', the diode is reverse biased and the current flows in the inductor and load, storing energy in the inductor. When the transistor is 'OFF', the energy in the inductor is discharged to the load through the diode.

In the steady state,

$$V_{out} = V_p d$$

where V_{out} is 'load voltage',

V_p is 'PV array voltage'

and 'd' is 'duty ratio', varied by pulse width modulation.

Assuming the converter to be loss-less, the supply power is equal to the load power,

$$V_p . I_p = V_{out} . I$$

where I_p and I are PV array current and load current respectively.

The PV array current is determined by measuring the voltage across the small series resistance. An analog voltage proportional to the power drawn from the array is generated by multiplying the array current and the array voltage. This voltage is converted into a digital number through an analog-to-digital converter and applied to processor-based controller. The controller generates a digital number which is converted into an analog voltage through a digital- to-analog converter. This voltage is compared with a ramp. Comparator output is applied to a series switch through a driver as pulse width modulation signal. At this pulse width, the load is supplied with certain amount of load current.

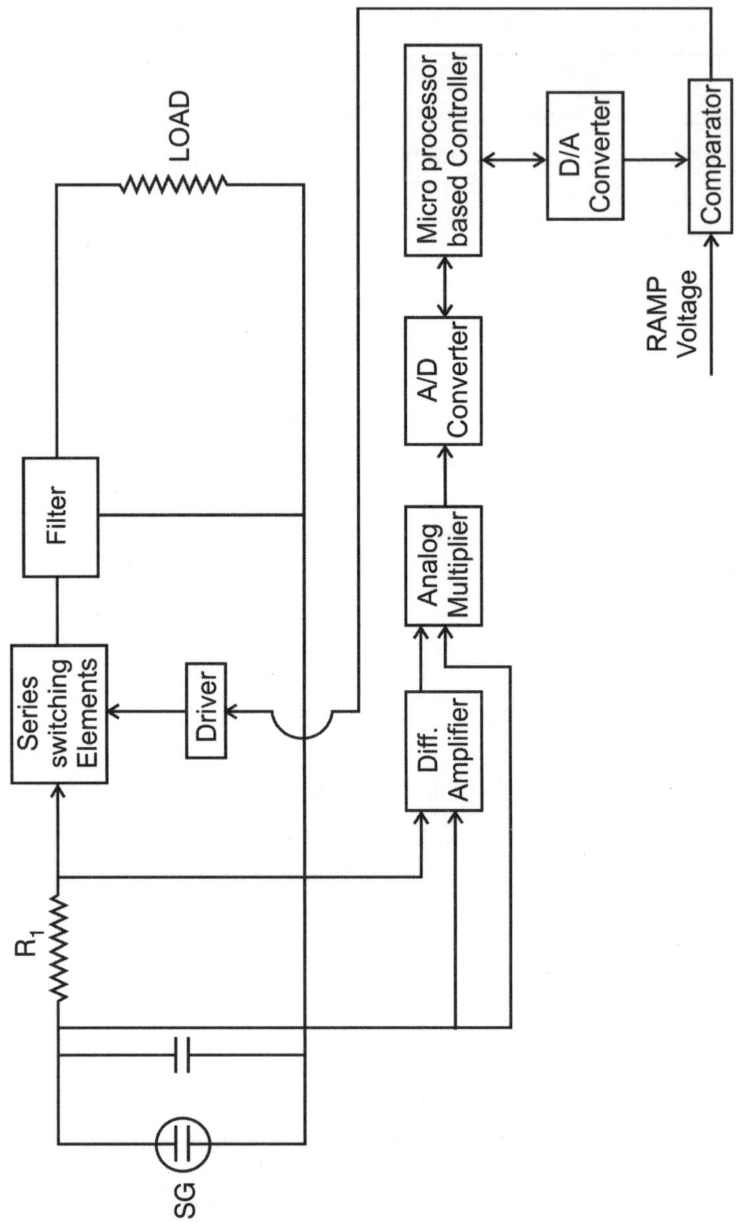

Fig. 6.5 *Block Diagram of typical Max. Power Point Tracking System*

The array power is computed and stored. As a next step in the sequence, the processor increments the digital number and array power is again computed. This array power is compared with the previous value. If the previous value is less, the number is further increased. Thus the digital number is sequentially increased until the array power starts reading. Thus the voltage for which the power output is maximum is obtained. In this mode of operation, maximum possible power will be obtained from the solar generator. The ordinary charge controllers only follow the I-V curve of the solar generator. The maximum achieved charging current is limited to the values of the I-V curve, whereas in the case of MPPT charge controller, with dropping battery voltage, the charging current will exceed the normal charge controller current.

MPPT is most effective under the following situations:

- **(a)** *cloudy days and winter days when extra power is needed most,*
- **(b)** *cold weather when solar PV modules work better, and*
- **(c)** *low battery charge conditions.*

The switching devices currently used are solid state devices like MOSFETs and IGBTs with very low ON- resistances and very good hf switching characteristics. Most of the functional blocks in Fig.6.5 are integrated into single application-specific microcontroller chips providing soft-ware controllable PWM outputs

Typical specifications of a charge controller are:

Input power:	1 kW
PV system voltage:	24 volts
Input voltage range:	V_p min. to V_{oc} max.
Operating mode:	Maximum power point tracking
Efficiency:	> 97%
Status indication:	Battery OK Battery low

Instrumentation:	Input voltage Input current Output voltage Output current
Protections:	
Input:	Circuit breaker or fuse
Output:	Current limiting circuit breaker or fuse

6.2 INVERTER

Inverter in a power conditioning subsystem converts the DC power generated by the PV array into AC at a frequency and voltage that are compatible with consumer utilities which run on conventional grid supply.

The main element in the implementation of a typical DC-to-AC is the power switching bridge. Fig.6.6 shows a typical full bridge illustrating the switching operation for conversion of input DC into AC.

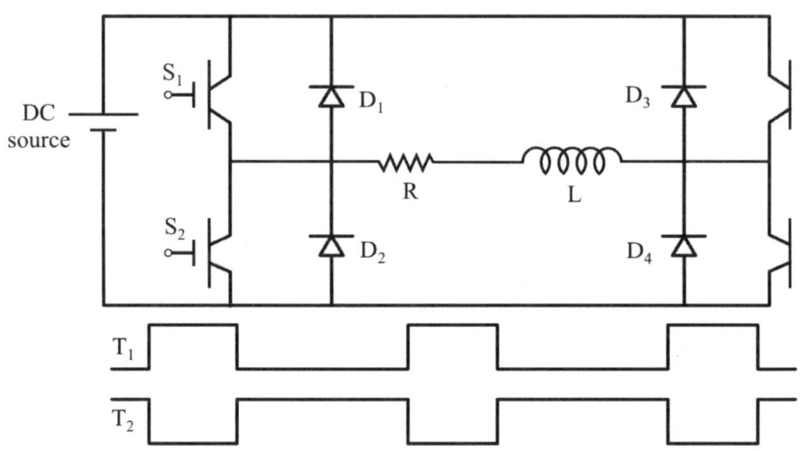

Fig. 6.6 *Full bridge switching configuration.*

S1, S2, S3, S4 are IGBT switches. S1 and S4 are driven to conduction by switching waveform, T1. During the positive half cycle, S1 and S4 are in conduction; the current from DC source flows through S1 and S4 through the load R,L. During the second half cycle, switches S2 and S3 conduct, while S1 and S4 are in off-state, the current from DC source flows through S3, the load and S2, reversing the voltage polarity across the load. Diodes D1, D2, D3 and D4 provide the necessary feedback path for energy stored in the inductor during the previous half cycle.

A generalized power conversion system is shown in Fig. 6.7 with a switched mode power converter. Normally, the requirement is for DC to AC inversion for most of the conventional loads. Based on the application, inverters could be square wave, quasi square wave or pure sine wave. Inverter technologies employ a wide variety of devices like power transistors, thyristors, MOSFETS, and IGBTs for design of power stages depending on the output power.

The important features of an inverter in a solar PV system are the reliability and efficiency.

The main performance parameters of an inverter that influence its design are:

Nature of application,

Wave form,

Conversion efficiency,

Input voltage and range,

Output power,

Output voltage and stability,

Power factor,

Frequency and frequency stability,

Harmonic distortion,

Self consumption,

Special load requirements,

Protection needed,

Safety requirements and Instrumentation.

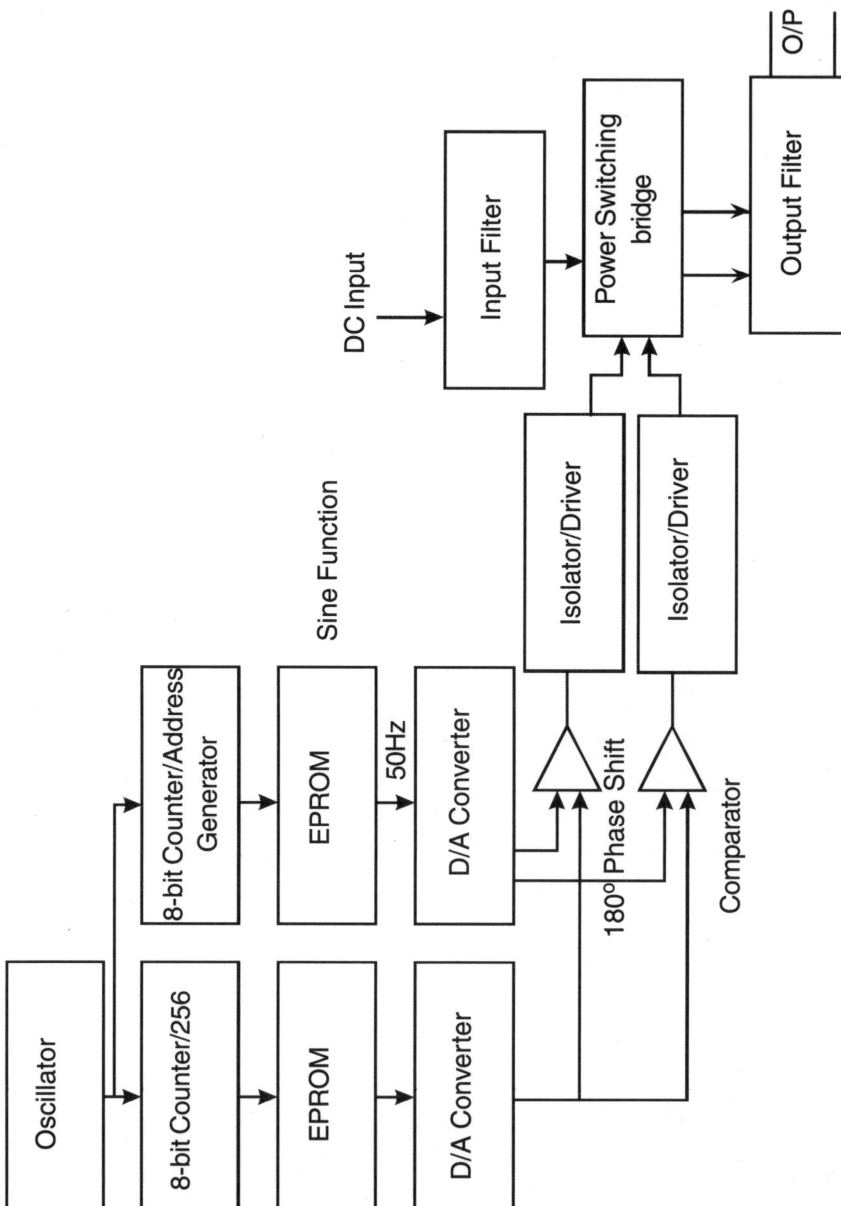

***Fig. 6.7** Sine wave Inverter for Stand-alone Applications*

Pure sine wave inverters are not cost effective for stand-alone PV systems used in remote site homes. Square wave inverters are inferior for any application except for resistive loads. Most stand-alone inverters approximate to a sine wave, generally referred to as 'quasi sine wave inverters'. Inverter causes interference in circuit electronics if the wave form is not clean. For example, if an inverter is powering both reactive and non-reactive loads, the reactive load changes the wave form of the inverter, giving rise to interference in the other load. An inverter must be capable of compensating for reactive loads and shall not distort the waveform.

Several different inverter concepts are used in installing a PV System. They are shown in Fig. 6.8 (Ref. 1)

(i) *a number of parallel PV modules strings connected to **central inverters**, which can convert the DC power from PV Generator upto several hundred kilowatts,*

(ii) *only one PV string connected to a **string inverter**, which can convert power between 500 Watts and 2.5KW, and*

(iii) ***Module Inverter** connected to a single module to the grid. The module inverters operates well even if some modules are shaded or if they are not of the same capacity. The inverter efficiencies, however, decrease with smaller systems.*

* The features and specifications of typical High frequency Sine Wave Inverters for different applications are given below.

6.2.1 High Frequency Compact Sine wave Inverter

The main features are:

* True sine wave
* Low Self consumption
* High Efficiency
* Short-circuit Proof
* Under and Over voltage protection

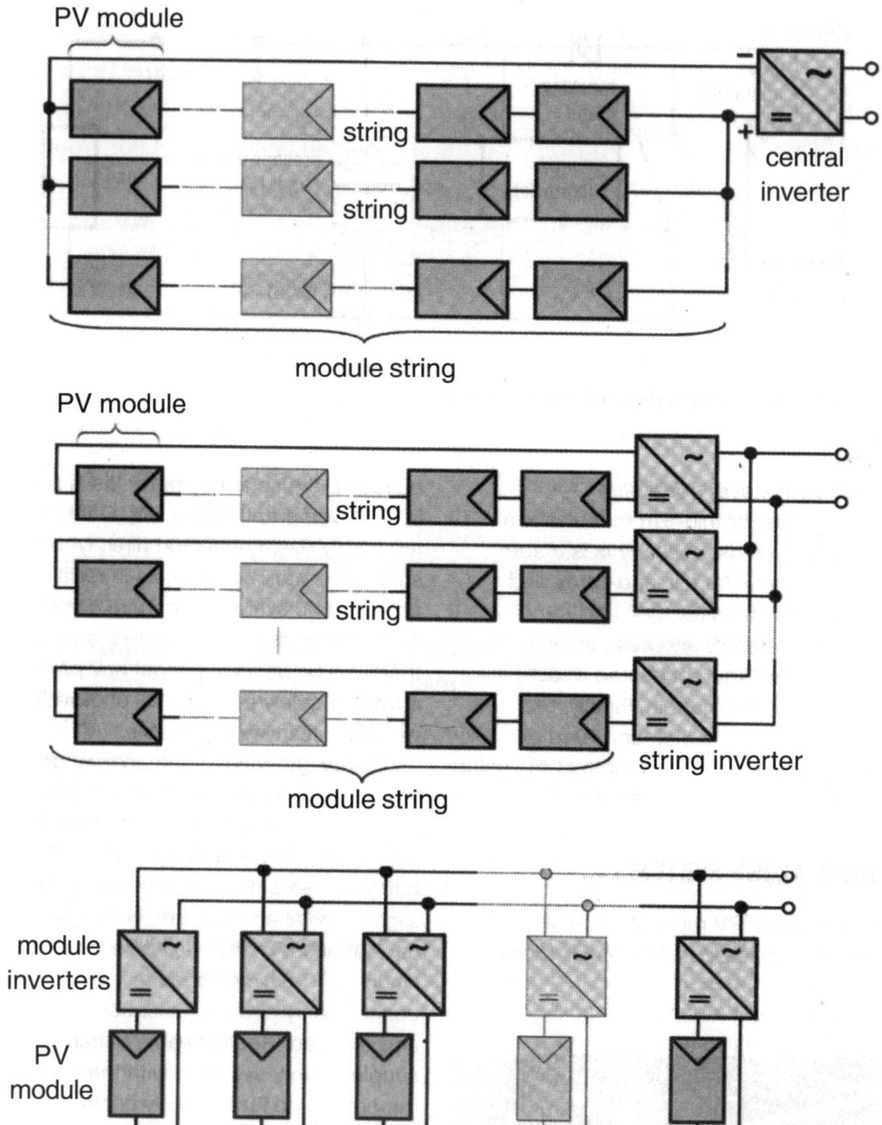

Fig. 6.8 *Connection of Photovoltaic modules to inverters using* ***(a)*** *Central inverters,* ***(b)*** *String inverters and* ***(c)*** *Module inverters*

* Overheating Protection
* Status indication

The specifications are:

Output power:	3 KVA, Single Phase
Input voltage:	48 volts
Output voltage:	220 V AC, Single phase
Output stability:	3%
Frequency:	50 Hz
Frequency stability:	0.5%
Waveform:	sinusoidal voltage
Harmonic distortion:	1....2%
Self-consumption:	1% of nominal power
Input Soft start:	Optional
Input protections:	Under-voltage and over-voltage
Output protections:	short-circuit proof
Instrumentation:	*input voltage *input current *output voltage *output current
Status Indication:	*input ok *output ok *inverter on *fault

6.2.2 Grid Connected Sine Wave Inverter

The features are:

* True sine wave
* Low self consumption
* Very high efficiency
* Lockout at grid under voltage

* Lockout at grid over voltage
* Lockout at supply under voltage
* Lockout at supply over voltage

The specifications are:

Output Power:	1 KVA, Single Phase
Input voltage:	24 volts
Input voltage range:	V_p min.....V_{oc} max.
Operating Modes:	stand alone output current control voltage tracking maximum power point tracking
Output voltage:	220 V AC, Single phase
Power factor:	> 0.95
Frequency:	50 Hz
Waveform:	sinusoidal
Harmonic distortion:	1 to 2% in stand alone mode 3 to 5% in grid-connected mode
Self consumption:	1% of normal power
Input soft start:	Optional for supply by battery
Output soft start:	standard
Input protection:	under voltage and over voltage
Output protection:	under voltage and over voltage
Instrumentation:	*input voltage *input current *output voltage *output current
Status Indication:	*input ok *grid ok *inverter on *fault

6.2.3 Three-Phase Sine Wave Inverter For Pumping Applications

The features are:

* True sine wave
* Low self consumption
* Very high frequency
* Dry running protection
* Turn-on procedure adjustable

The specifications are:

Output Power:	2.5 KVA
Input Voltage:	240 V DC (maximum power point)
Input voltage range:	210 V(V_p min.) .....360 V(V_{oc} max.)
Operating modes:	maximum power point tracking voltage tracking battery supplied
Output voltage:	220V AC/ 380V AC step-up transformer
Efficiency:	over 10...100% load 90 - 95%
Frequency:	0....50 Hz
Frequency limits:	Upper frequency limit adjustable Lower frequency limit adjustable
Ramping conditions:	Adjustable
Wave form:	sinusoidal voltage
Self consumption:	25 W approx.
Output protection:	current limit circuit breaker
Input protection:	circuit breaker
Lighting protection:	optional

Dry Running Protection signal input provided.

6.2.4 Efficiency of Inverter

In the case of solar PV power generating system where the maximum energy has to be derived from the PV resource the conversion efficiency of the inverter becomes very important. With recent advances in solid state switching device technology, efficiencies of the order of 97% and above have become very common. It is reported recently that efficiencies up to 98.5% have been achieved using SiC MOSFETS. Inverter efficiencies are generally specified either as maximum or as a range over normalized load. Typical inverter efficiency curve is given in Fig.6.9. A maximum efficiency of 97.5%, and 90-95% over 10-100% normalized load are seen.

European Union defines 'efficiency' as a weighted average over the normalized load range, 5 to 100%, as (Ref.2)

$$ç = 0.03\ ç_5 + 0.06\ ç_{10} + 0.13\ ç_{20} + 0.1\ ç_{30} + 0.48\ ç_{50} + 0.2ç_{100}$$

For the efficiency profile shown in figure, European efficiency works out to 95.8%.

While considering the efficiency of a PV inverter, other factors like the input voltage range and temperature range of operation that affect the efficiency need to be considered.

6.3 TEST INSTRUMENTATION FOR BOS

'Solar Generator Simulator' simulates array characteristics to enable testing of balance of system (BOS) components such as charge controller and inverter. The features of Solar Generator Simulator are:

* Selection of different I-V curves
* Preset of Open circuit voltage
* Preset of Short circuit current
* Simulation of PV array power profile from sunrise to sunset
* Simulation of I-V characteristics under different cloud conditions
* External or manual control of I-V curve

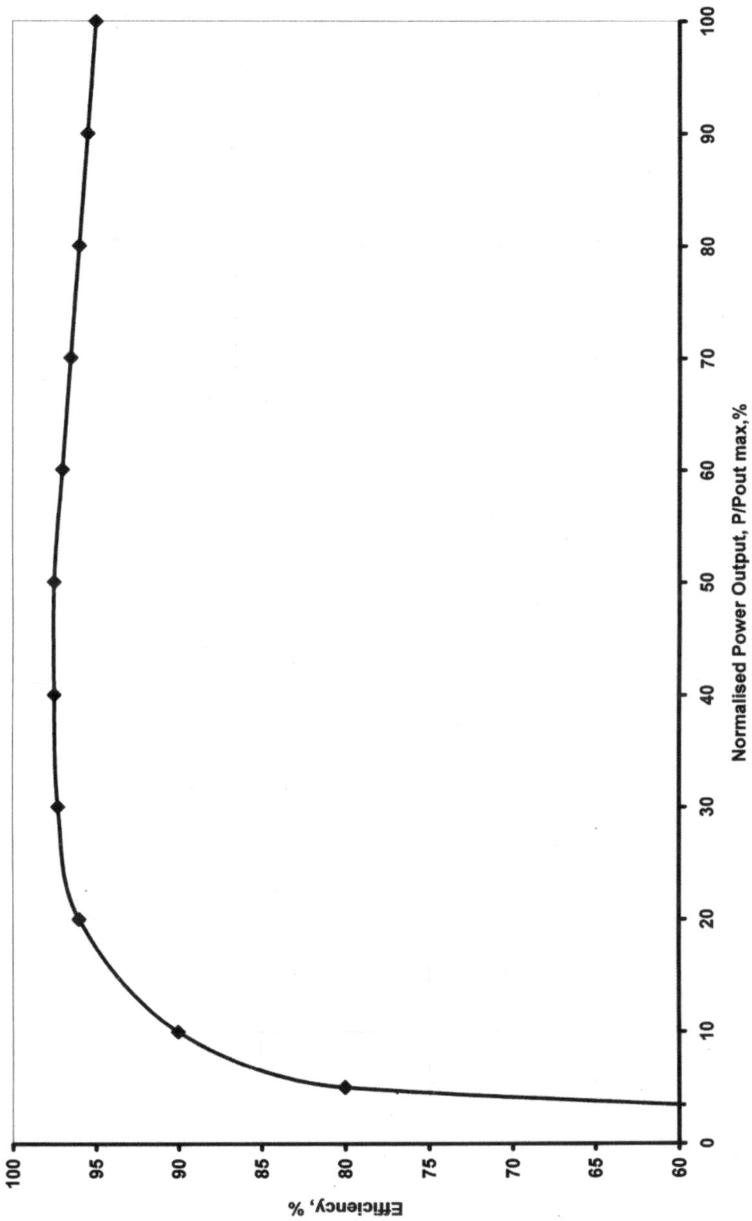

Fig. 6.9 *Inverter efficiency curve.*

REFERENCES

1. ***V. Quaschning:*** *Renewable Energy World, 7, p81, 2004*
2. ***Haeberlin et al:*** *13th European PVSEC, Nice,1995*
3. *For recent designs in Inverter technologies and applications, refer to 'Solar & Wind Energy' issues published in 2007, 2008, 2009.*

SUGGESTED READING

1. ***Jacob J. Michel:*** *"Power Electronics: Principles and Applications", Albany: Delmar Thomson Learning, C2002, xvii, p525.*
2. ***M.H. Rashid:*** *"Power Electronics - Circuits, Devices and Applications" (3rd Edition) upper saddle River, N.J. Pearson Education (2003) (Also Prentice-Hall).*

CHAPTER SEVEN

DESIGN, INSTALLATION AND OPERATION OF A SOLAR PV SYSTEM

7

DESIGN, INSTALLATION AND OPERATION OF A SOLAR PV SYSTEM

7.1 INTRODUCTION

The designing of a customized solar photovoltaic system is simple, provided the required information is obtained precisely. For instance, predicting solar intensity and weather conditions daily, monthly and annually at the site of installation of a SPV system is a difficult process. But this is a major input in the design of the system. In the absence of such precise data, the average daily solar radiation at the place or at a place nearby can be obtained from meteorological sources. Such averages are calculated using solar radiation data collected over the past few years. But these averages are likely to change year after year. Moreover, there is no guarantee that a system designed on the basis of such a daily average insolation value for the month, will perform satisfactorily for all days of that month and for the entire year. One has to, therefore, depend on statistical approaches to estimate the probability of occurence of a certain insolation value on a certain day. Therefore, a suitable combination of statistical and engineering aspects are needed for a good system design.

It is also difficult to predict the load with any accuracy because the consumption patterns vary tremendously over the year. To assess the number of photovoltaic modules required and to calculate the battery size, the daily energy requirements and the solar insolation data are primary and also critical. This is particularly true for stand-alone PV installations which have no backup power sources.

The issues to be examined in the process of system design are many and some of them are complex. Some manufacturers of PV modules, batteries and charge controllers advocate that once the load particulars are known, it is easy to design the system using certain suggested procedures, referring to some charts and tables provided by the manufacturers. In practice, this approach has proved disastrous, because neither the PV modules nor the batteries nor the inverters work with 100% efficiency. The PV array's output is decided by the solar insolation and the ambient air temperature; the batteries in the conversion of electrical energy to chemical energy to electrical energy are only 80 to 90% efficient, and the inverters have invariably a 15% loss of energy for any AC load. The sunshine hours at a place vary with the seasons of the year; the days are long and the nights are short in summer and vice-versa in winter. Consequently, the power output from a PV array is more in summer months and less in winter months; although hot temperatures in summer months have the adverse effect on the output from the solar modules. At the same time, longer nights in winter result in increased loads, requiring more power. These factors have to be properly balanced and optimized in designing a PV system, to achieve desirable performance with cost viability.

7.2 CONCEPTUAL ASPECTS OF SYSTEM DESIGN

The statistical approaches are inevitable in the design of Solar PV systems. The system reliability can be predicted with reasonable precision, if the statistical variations of the inputs, such as the global and diffuse components of solar insolation, the declination of the sun, the ambient temperature, the sunset hour angles are properly estimated. The correct assessment of the insolation and other related weather parameters are possible if the necessary statistical deviations are introduced by means of suitable standard deviations. The estimated weather data may be based on annual or monthly or daily or hourly values. The reliability of the system however depends on duration of the year is studied for these values and the type of statistical deviations.

The system losses, the load variations and any degradation of the system components over the designed life also need to be assessed for arriving at correct system specifications. These inputs along with a set of assumptions form the basis of the iterative calculations to arrive at an optimum design of the system. The assumptions are:

(i) *the system design is highly site specific,*

(ii) *the system has to work satisfactorily, particularly when there is no backup source,*

(iii) *the performance and reliability of the system should continue to be the same for the entire life of the system,*

(iv) *the energy generated by the system must be at least be equal to the total load plus the system losses at any point of time,*

(v) *the batteries cannot be overcharged; at the same time it cannot be discharged below the maximum DOD specified by the manufacturer,*

(vi) *a particular combination of SPV array-Battery bank that meets the system specifications should also match the cost-economics and minimization of energy generated by the system due to the fully-charged condition of the battery.*

A system design is considered to be optimal if the PV array, battery bank, charge controller, inverter and other subsystems meet the system specifications satisfactorily at the acceptable cost, and provide reliable performance during its designed lifetime.

7.3 DESIGN PARAMETERS

The optimum design of a SPV system involves examining several parameters and the most important ones are the following:

(a) Equivalent Hours of Full Sunlight (EHFS):

It is the solar energy falling in one day on a flat surface of one square meter area placed on earth's surface. it is expressed in hours per day at an intensity of 1000 watts/sq.meter which is equivalent to 'one sun'. The EHES is quite different from the number of sunshine hours which are generally referred in the meteorological data. The concept of EHES is convenient to estimate the total energy output from the system, although it is an 'average' value for the day or month. The magnitude depends on the inclination of the flat surface on earth as well as whether the mount is static or tracking type. The EHES is also influenced by the geographical, weather and other environmental conditions of the place.

(b) Optimum Array Tilt Angle:

Since the incident solar energy on a PV array depends on the direction of the direct sun's radiation, the efficiency of the System can be enhanced if maximum incident solar radiation on the array is ensured. The ideal position for an array to get maximum exposure to sun is when the sun's radiation hits the surface of the array perpendicularly. Hence, the maximization of incident solar energy can be possible if the array is made to track the motion of the sun throughout the day. A two-axis tracking mechanism can ensure this feature of the design, though it increases the cost of the system.

In practice, however, a fixed array facing South (in the northern hemisphere) and oriented at an angle equal to the latitude of the place, is found to receive maximum solar radiation, irrespective of System specifications. But as the seasons of the year change, the array tilt angles other than the latitude of the place, have to be considered for the maximization of the incident solar radiation. The best tilt angle is obtained by adding 15 degrees to the latitude in the winter season, subtracting 15 degrees for the summer season, and maintaining at latitude in the spring and autumn seasons of the year. These array positions can be adjusted manually. It is also viewed that the 'fixed' mount is a good alternative to varying the tilt angle due to maintenance problem, though the system size has to be increased to compensate for the reduced efficiency. The best tilt angle for the 'fixed' mount is 'latitude plus 15 degree for the entire year because this maximizes the winter gain and the reduced summer output is compensated by the increased sunny days and decreased lighting load.

It is also observed by experience that a perfectly horizontal-fixed-array is most suited to places at/or near the Equator, though this practice is not generally followed.

(c) Loss-of-Load Probability (LOLP):

The LOLP essentially indicates the percentage of time in a year that the SPV system is likely to be shut down, and it is an indicator of system reliability. It is also expressed as number of days per year during which the System is not operative as per the design. One of the reasons for this situation could be 'No-Sun days', when direct sun's radiation is not falling on the array. LOLP can never be zero, which means that the system delivers power all the 24-hours-a-day and all the days in a year. Such a design will be highly uneconomical. How ever, for stand-alone SPV systems, it is an important design parameter; and a balancing of the reliability and the cost are essential for an efficient operation of the system.

7.4 SYSTEM DESIGN PREREQUISITES

The sizing of a PV array is only one aspect of the system design. There are a few requisites to be complied with. Firstly, to decide the system size in terms of the AC output power it delivers; secondly, to find the roof or ground surface that faces south at the site of installation, free from obstacles and from shadows cast by the nearby structures or trees; thirdly, to select a power conditioning system to match the array design based on its ac power level, dc input voltage, efficiency and cost; fourthly, to examine the PV array/power conditioning system interface issue which influences the safety of System configuration.

7.5 ESTIMATION OF LOAD

To accurately size a PV system, a correct assessment of the load is essential. Since the consumption pattern of power vary with the season of the year, the daily average load has to be determined. A list of electrical gadgets in a typical home along with the wattage of each gadget is prepared. The format used to estimate the load in watt-hours is given in Annexure 5. Due to changes in the load pattern during the year, one has to prepare for the worst case, namely winter, when the solar insolation available is least and the load is high.

In some instance, the worst case for the load may be the summer and the worst case for the insolation is winter. The excess load may be served by an auxiliary power source. If it is a stand-alone type, two separate designs are worked out and the one that will carry the load through both the seasons is selected which will be uneconomical. Alternately, the PV system capacity is averaged by working out a 'daily average load'. In deciding the daily average load, the losses due to inverter, the battery, the wiring and the interconnections are included. To compensate for these losses, some designers multiply the 'estimated load' by a factor of 1.5. In most designs about 15% is considered for inverter losses and 20% for battery losses, and the 'load' is accordingly corrected. There is no single defined practice in accounting for these losses. Nonetheless, a thorough and realistic assessment of load is very important because most of the 'system' failures are the result of 'underestimated' loads.

The seasonal average peak sunshine hours/day for a particular location are obtained from meteorological observations. The figures are averages for a season, ie, a 3-month period, and monthly differences can be significant. These figures may also vary from one year to the other. This data is season- and location- specific also. It needs to be emphasized that obtaining reliable data is mandatory for the successful designing of the system.

7.6 SIZING OF THE ARRAY

The size of the PV array is decided largely by the 'average daily peak sun hours' available at the place of installation for each of the four seasons, Summer, Fall, Winter and Spring. 'The International Solar Irradiation Data Base' provides insolation levels for various regions of the Globe (Ref. 2a, b).

The estimation of ' the average daily module output' from the PV modules selected for the system is performed as follows:

Step 1: *The 'average daily peak sun hours' for the four seasons are taken from the weather-data for the location of System installation from the Annexure 6. The 'fixed yearly average daily peak sun hours' is estimated by taking the average of the values for Fall and Spring seasons.*

Step 2: *The 'average daily usable module output' in watt-hours are then determined for the four seasons. To obtain these values, the PV modules to be used in the proposed System have to be decided from the data given by the manufacturers. The 'usable module output' for these modules are then estimated. For example, assume that 75 watt modules of a particular manufacturer are selected. 75 watts is the 'rated capacity' (4.4 amps at 17 volts) for a typical load as furnished by the manufacturer. But its rated output has to be derated to get the output for normal operating voltage of the battery system, namely 12 volts. ie, 4.4 amps is stored in the battery which supplies to load at 12 volts, which means that 4.4 amps x 12 volts= 52.8 watts only is the 'usable module output' from the 75 watt module. The product of 'usable module output' and 'the average daily peak sun hours' gives 'the average daily usable module output' in watt-hours.*

Step 3: *The average daily usable module output' is then estimated for all the four seasons, by multiplying 'the usable module output' by 'the average daily peak sunhours' of the respective seasons. At this point, the designer shall decide whether the array will be 'fixed' or 'adjustable'. If the array is fixed one, then a tilt angle of (latitude + 15 degrees) is chosen for the best possible output, even for the worst season, namely winter of the year. The 'yearly average output of the fixed array' is estimated by taking the average of the 'fall' and 'spring' outputs for an angle equal to the latitude of the place. This approximation is considered to be near realistic for the 'fixed' array system.*

Step 4: *The number of modules needed to the System are then estimated by dividing 'the average daily load' by 'the average daily usable module output'. The number modules needed therefore vary from season to season. For simplicity, however, the PV system can be sized for the yearly average daily output of a fixed array, fixed at the winter array tilt angle.*

These calculations can be carried out for other array tilt angles also, if the system design requires the season-wise adjustability of the tilt of the array.

The 'format' for System sizing is given in Annexure 6.

In selecting the modules, the following aspects of the modules have to be studied:

(i) *The technical specifications given by the manufacturer along with the warranties, the reliability and the cost,*

(ii) *The electrical configuration of the solar cells in the module decides whether the module is a high voltage-low current one or low voltage-high current one. The type to be chosen is decided by the power conditioning system's input voltage and the desired ac power of the system.*

(iii) *The 'peak ratings' of the modules furnished by the manufacturer have to be converted into 'site-specific' conditions, also called 'standard operating conditions' to achieve a realistic design.*

(iv) *The electrical configuration of the array which helps to achieve maximum electrical reliability.*

7.7 SIZING OF THE BATTERY SYSTEM

Battery sizing is a ticklish problem because design parameters such as LOLP and EHFS influence the exercise. Normally the size of the battery in amp-hrs(AH) is 2 to 5 times the PV array's power in peak-watts.

But there are a few factors, namely, compensating for bad weather (no sunshine days) and maintaining 30% reserve in the battery after a deep discharge, to be considered in deciding the battery capacity. These factors in turn push the cost of the battery bank upwards.

Most batteries serve longer if they are shallow cycled; ie, discharged by 20% of the capacity only, instead of deep-cycled; ie, discharged by 80% of the capacity as a regular feature. The simplest and most practical method of estimating the capacity is outlined:

Step 1: *The daily output of the PV array in the System in watt-hours is divided by the voltage rating of the battery(6 volts or 12 volts), to get the battery capacity in AH,*

Step 2: *To provide the autonomy for the battery, the capacity in AH obtained in step 1, is multiplied by the maximum number of days of bad weather to be compensated. In general, 3 to 10 days are considered; but it must be noted that more the number of days, more the battery capacity and higher the cost of the system.*

Step 3: *The resulting figure in step 2, is divided by 0.7 to account for the 30% reserve in the battery after a deep discharge. The value arrived at is 'the total battery storage capacity' to be installed in the PV system.*

To balance the cost and the battery size, it is suggested to utilize a 'back-up' power source like a generator or grid supply, for the worst period of the year.

7.8 SELECTION OF THE ARRAY MOUNTING

The mounting of the array is an important aspect in the design, installation and operation of the PV System, because the mounting scheme influences the cell operating temperature and in turn the efficiency. Small PV systems used for domestic applications are 'roof-mounted' whereas bigger systems are installed on the ground. The nature of the mechanical supporting structure used to fix the PV array decides the type of the 'array mounting'. In practice, there are four types of 'mounting schemes', each one possessing specific features (Ref.3) by which its suitability for a particular location is decided.

(a) Rack Mounting:

It is a simple mounting design which has a flexibility for fixing the orientation and tilt angle. The designer can reach the array wiring on the rear side easily. This mounting is considered as the most suitable configuration for the installation on a flat roof. The modules are attached to wooden frames mounted on the rack and fixed to the flat roof using aluminum brackets. This mounting involves additional cost and is not suitable to non-flat roofs.

(b) Direct Mounting:

In this configuration, the modules are directly attached to the roof sheathing without a mechanical support structure; there is saving of the cost of the structure and part of the roof-laying expenditure. The installation has to be planned along with the house construction. Here the array lies very close to the roof allowing limited space for the electrical configuration and the wiring. In recent years, the direct mounting has become popular due to the availability modules appropriate to this mounting scheme.

The main disadvantage with this mount is that the array operates at high temperatures reducing the conversion efficiency. This may be partly countered by arranging heat dissipating mechanism at the rear of the array structure.

(c) Stand-off Mounting:

This configuration is almost similar to 'the rack mounting', but with reduced support structure costs and better appearance. The array is arranged parallel to the roof surface of the building. The accessibility to the rear of the array and the wiring is, of course, limited, though it is much preferred mounting arrangement for residential buildings. The modules are arranged using roof jacks along with holding frames for providing structural strength and rigidity and are electrically connected. This mounting scheme is highly suitable to both retrofit and new buildings.

(d) Integral Mounting:

In this mounting scheme, the modules themselves form the roof,- the modules being supported by specially designed structural framing. The cost of the roofing material is highly reduced in this scheme.

In recent years, the building-integrated- photovoltaics are largely employed in Europe, north America and Australia. The electrical interconnections and wiring are easily accessible on the rear enabling installation and servicing easy.

The selection of the mounting scheme for a location is decided, considering the following factors: (i) whether the system will be integrated into a new construction or an existing building, (ii) the nature of roof construction and the roof orientation, (iii) the climatic and environmental conditions of the location and (iv) the array's electrical performance.

Each mounting option creates special design constraints in the PV System design.

7.9 SAFETY AND PROTECTION

The photovoltaic power generating system, like other power plants, requires proper protection and people who work with have to follow certain safety measures. The concern for safety measures is much more in PV power systems because they continuously supply electric power and cannot be switched off. Even full moon light illumination will produce small voltage.

A 5 kw residential PV array generates roughly 25 amps of current under full solar illumination, and about one milli-ampere under full moon illumination (Ref.4).

The PV safety standards exist in all countries and these standards have to be maintained right from wiring the modules to installation of the System. All the circuits in the PV system need to be insulated, and the conducting frames grounded as per the code provided. The risks involved due to electric shock and fire are primarily due to insulation failure; it is mandatory to use insulated materials which can withstand the environmental stresses.

Wiring, grounding, selection of fuses and breakers, sizing conductors, locating disconnects, diodes and fuses are the most important jobs to be carried out for the overall safe functioning of a PV power system. In essence, good engineering practices and adherence to codes and standards are essential.

(a) Fusing:

The basic PV system consists of a PV array, a charge controller, a battery bank, an inverter and wiring. The line resistance is very critical in the array wiring, because it results in power losses. This problem is more serious in systems at low voltages and high currents calling for strict implementation of the code in choosing the wiring. The operating current is generally higher than the short-circuit current by about 15%; hence the conductor size should be at least 15% higher than the recommended value. Incidentally, such enhanced conductor size would help to handle higher currents resulting occasionally for short durations due to increased solar insolation.

Fusing is another important criterion for safety. Fuses or current breakers are included at different places in the System for over current protection and for other reasons. A fuse is placed between the battery and the load (dc load) to break the circuit in case a short occurs. It is possible for the current to flow back from the battery to the charge controller if the modules or the controller or its wiring directly short; a circuit breaker or a fuse between the charge controller and the battery bank avoids this danger. This fuse is very important. The lines from the battery to the inverter shall also include a dc current breaker or fuse. And the inverter shall have its own output fuse protection in the form of a breaker in the Service Panel. If the inverter is self-current limiting, a melting internal type fuse is desirable.

Fuses and circuit breakers are either AC, DC or AC-DC type and are rated for amperage and voltage. A dc switch or circuit breaker utilizes heavy duty contacts whereas ac switch/circuit breaker utilizes lighter duty ones .The dc rated switches or circuit breakers are more expensive,- fuses are preferred to keep the costs low. The fuse can be removed to disconnect a circuit. Getting a right kind of fuse for the right place in the circuit is very important. In addition to fusing, there should be a facility to disconnect one section of the system from the other to enable installation, repair and servicing and for any other emergency. The system designer provides a switching fuse box or a circuit breaker or an add-on switch at every point of over current protection in the circuit (Fig. 7.1).

(b) Grounding:

Like fusing, grounding is very important in a PV system. It provides a path of low-resistance from various points in the system to earth to carry current in abnormal situations. A PV array is a weak source of voltage; and during a fault, it can register a small increase in current. The array current cannot be switched off immediately, if the sun is shining; and an effective 'grounding' would help to avert the problem. The grounding is done for the entire PV system at the mounting structure. Frames for mounting the PV modules are generally made of iron or steel and to avoid electrical contact, washers(insulating separators) are provided between the modules and the frame. Stainless steel bolts are used as fasteners. The PV system is grounded at each module frame,

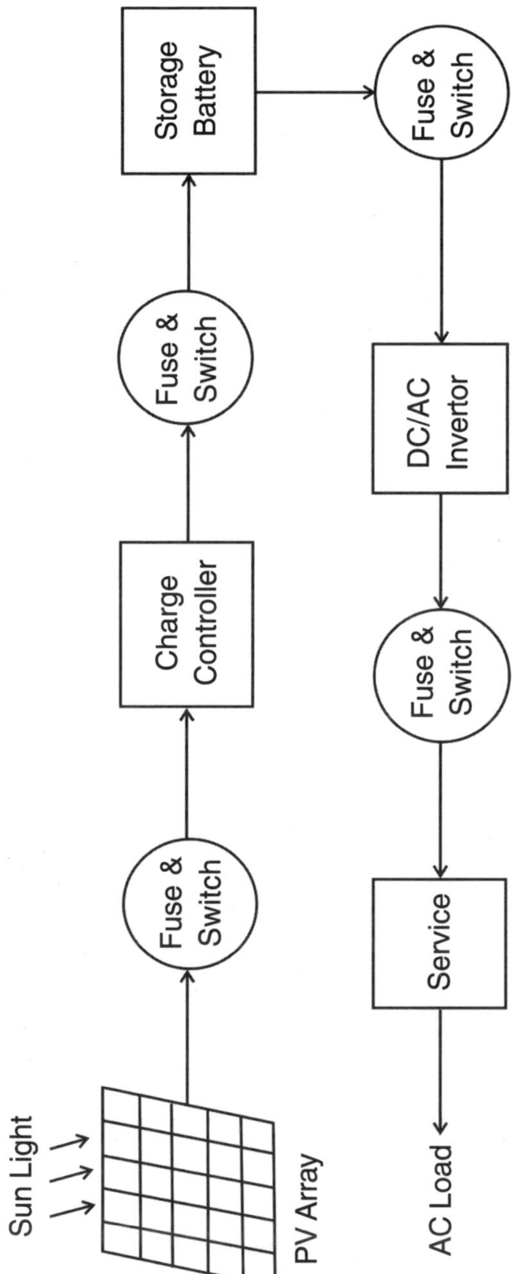

Fig. 7.1 *Location of Overcurrent protection points in a PV System*

at the negative end of the battery system, at the inverter casing and at the ground bus in the AC fuse box. Ground path is provided to control the surge currents which may arise from the array or from the internal switching of power or from the utility. A good grounding of the mounting structure saves the array from direct lightening strikes (surges). The largest wire with the shortest straight path is used for grounding to create least resistance and fast dissipation of lightening surge. The PV modules can resist the damage due to lightening, if the frames are properly grounded. Other elements such as diodes, sensing equipment and charge controllers need protection from 'surge' currents. Fast-acting voltage surge arrestors are used to protect inverters and other electronic controls connected to the array. The power surge can degrade the wiring insulation also. The array wiring automatically gets grounded when it is connected to the battery bank. The metal boxes of fusing, circuit breakers and charge controllers also need to be grounded. The inverter should be protected on the ac-side more than on the dc-side. Most inverters have surge suppressors to protect the circuitry from high voltage surges created by reactive loads.

If different parts of the system are connected to different grounds, then all grounds must be interconnected. The location of the grounding points are shown in Fig. 7.2, and the locations for the lightening protection are shown in Fig. 7.3.

(c) Physical Structure:

The positioning and fixing of the components of the PV system is very important. It is possible that the glare from the PV array may disturb adjacent buildings or public places or traffic areas. This must be avoided by properly positioning the array without sacrificing the efficiency of energy conversion. it is also necessary to ensure that the system maintenance and the routine house maintenance do not conflict. The accumulation of debris under the array, which generally happens, should be avoided.

The supporting structure for the battery system must be strong enough to support the weight of the batteries. If the structure is metallic, it is coated with a non-conductor which is resistant to damage from the acid. Lead-acid batteries

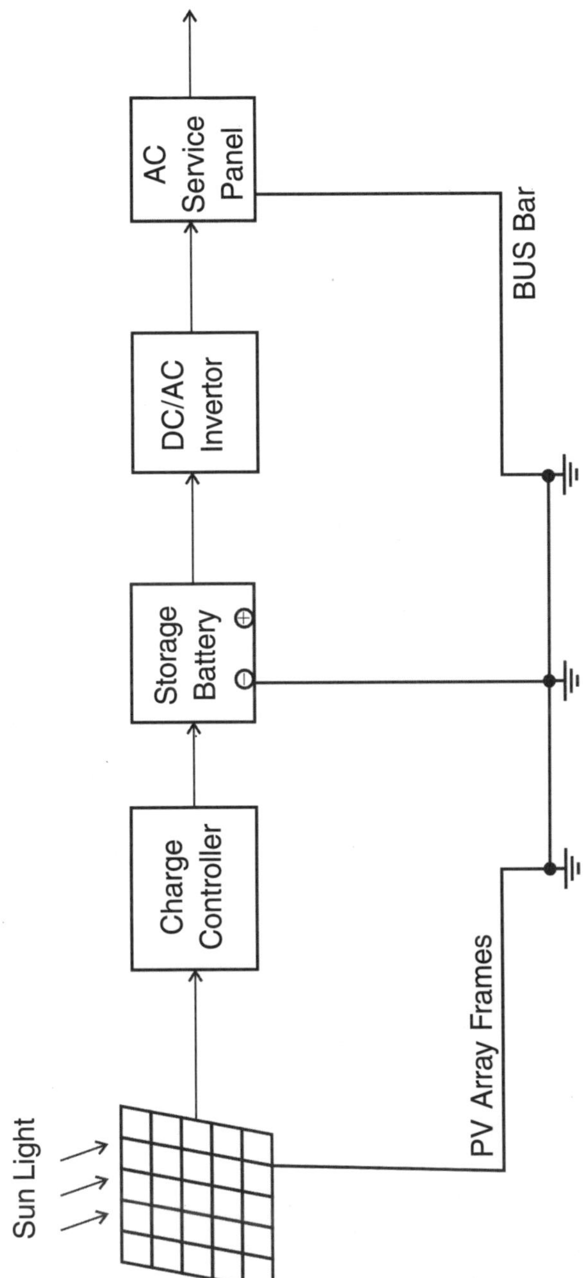

Fig. 7.2 *Location of Grounding points in a PV System*

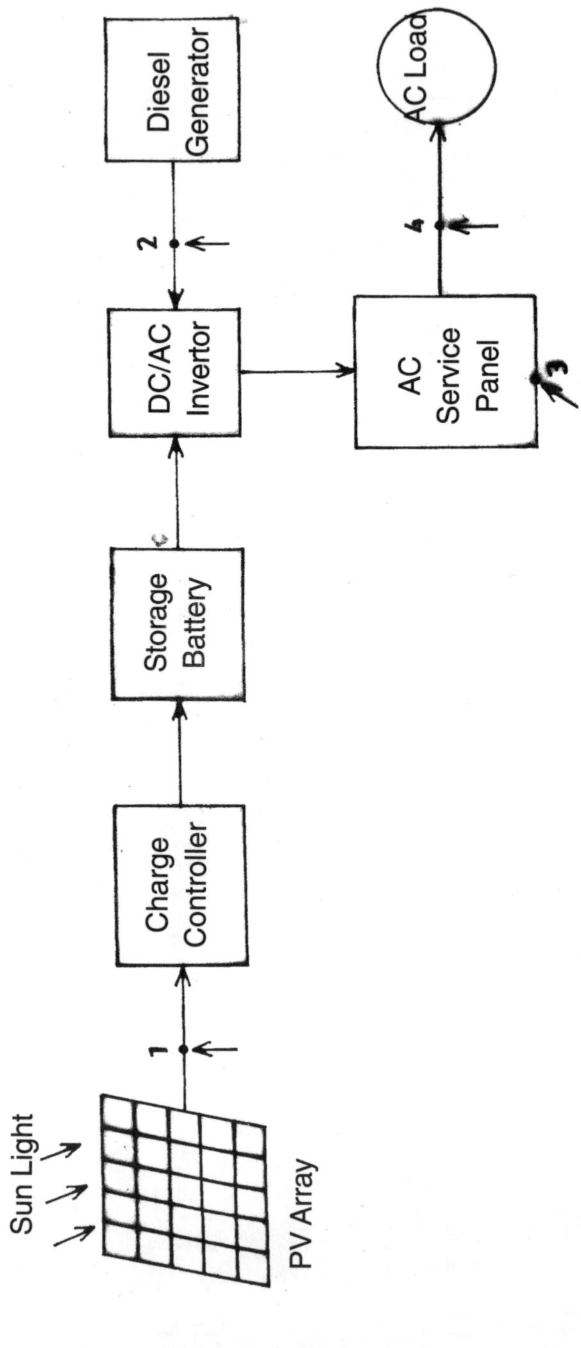

Fig. 7.3 *Lightening protection points (1, 2, 3, 4) in a PV System*

give off hydrogen and oxygen during charging. It is necessary to fix a venting system made of polyethylene, six to eight inches above the battery system for air circulation. The battery system, the inverter and other controls are installed in a cool well-ventilated area. The battery bank is separated by a wooden partition to shield it from the potential sparks from the charge controller, inverter, meters and fuse boxes. The presence of gases over the batteries also necessitates the avoidance of sparks and flames over them.

Sometimes electricity flows in the backward direction from the array due to faults or array string mismatch and damages the modules and increases energy losses. To prevent this reverse flow, 'blocking diodes' are fitted in the PV array. By-pass diodes are included to prevent hot-spot damage arising due to the shade falling on the modules. The diodes may be internal to the module or sometimes mounted externally.

Standard AC and DC wire as per the specifications/code is chosen; the wire sizes should be adequate to withstand the current strengths of the circuit, the over current protection and the length of the run.

A typical PV power generating system including safety and protection elements is shown in Fig. 7.4.

7.10 SYSTEM INSTALLATION

The installation, operation and maintenance of the system at a particular location is as important as the designing of the system to deliver the expected performance over its lifetime. It is essential to provide guidelines to the three groups,-the installers, the owners and the maintainers who have specific roles to play in the running of the system. The installer, the owner and the maintainer have to be familiarized with the description of the system to understand the guidelines and to implement properly. The principal components that constitute the system are PV array, source circuit, charge controller, battery bank (for stand-alone systems), and inverter. The other components are bypass diodes, blocking diodes, fuses & surge arrestors, source circuit disconnect & array disconnect, and AC disconnect (for grid-connected systems). The three groups should be familiar with the functioning of these components.

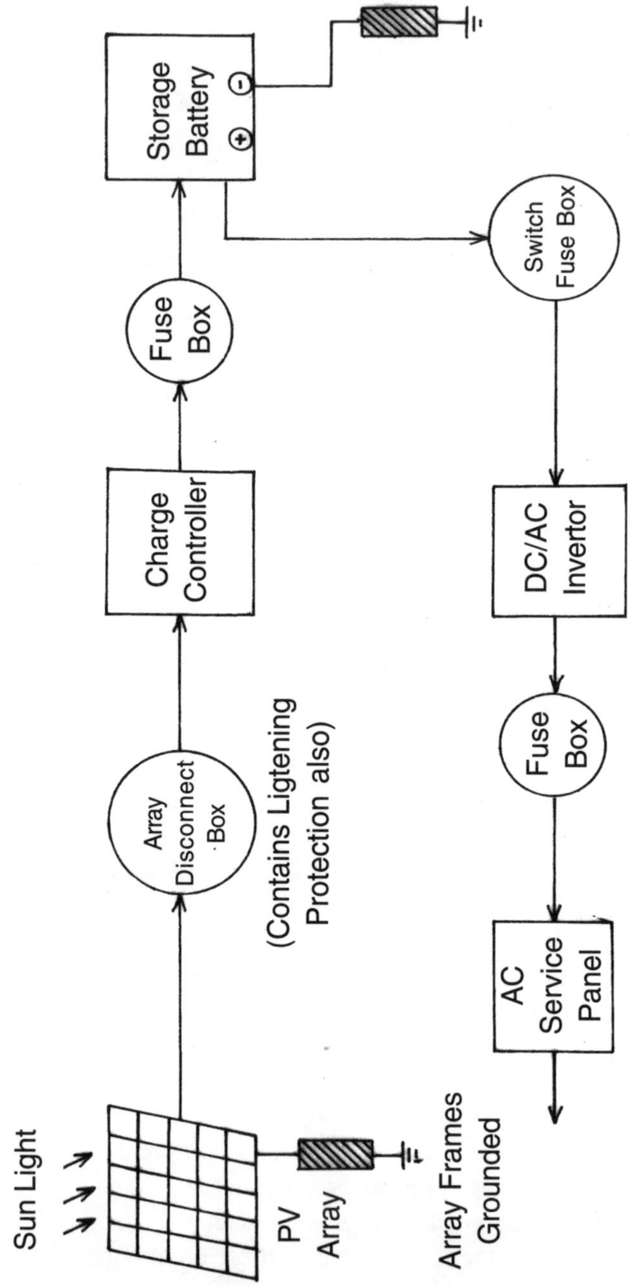

Fig. 7.4 *Typical Solar PV Power generating System*

INSTALLATION GUIDELINES

The installation personnel are technically trained to handle the electric wiring, testing of array and power electronics, the installation of array, electronic equipment and the protection devices, and testing of the total PV system.

After the system is ready , the installation is carried out as follows:

Step 1: *Modules should be visually inspected for any defects or damage such as broken cells and/or glass sheet, dents to frames etc; the charge controller and the inverter should also be examined for any damages,*

Step 2: *Modules are tested for their electrical performance by measuring the open-circuit voltage (V_{oc}) which should tally wih the manufacturer's specified value within 15 to 20%. This allowance is made because the weather conditions (temperature and intensity of solar radiation) at the time of testing may be different,*

Step 3: *The mounting scheme shall be decided depending on the construction design, the roof style and the local climatic and environmental conditions.*

The mounting method must allow easy physical accessibility to the electrical terminations at the rear side and also safe handling of modules and the wiring during servicing and maintenance. The modules are then installed on the mounting structure and the electrical interconnects are attached ,ie, the modules are connected in series to form a source-circuit to achieve the designed system voltage and then these source-circuits are paralleled with other source-circuits to form the complete array output.

For integral mounting in the case of new buildings, the 'building integrated PV modules" are installed during the construction itself. The installation procedures provided for the corresponding mount have to be followed. For instance, grounding the array-frame before the modules are mounted, checking open-circuit voltage on each source-circuit as the wiring is completed following detailed wiring procedures, and 'not connecting' the array to power electronics before the system is completely done have to be scrupulously followed.

Step 4: *Location of power conditioning equipment depending on their physical characteristics, dimensions and operational requirements(the operating temperature and humidity) as given by the manufacturer. It must be located in protected area to avoid personnel coming into contact with active electrical elements.*

Step 5: *Installation of System protection devices,- over current protection devices, surge protectors, manually operable switches, circuit breakers and feedback protection from any source supply.*

Step 6: *Locating PV system components such as fuses, blocking diodes, source-circuit disconnects in the dc service box to protect service personnel and to facilitate maintenance.*

Step 7: *Labelling of warnings and special instructions must include ac/dc disconnect points, system circuit, test point labels and approximate readings, possible voltage ranges at junctions etc.*

Step 8: *Final inspection of the total PV system for its structural safety and electrical performance.*

In all these steps, the installation personnel shall take all precautions to ensure a proper installation, a safe and secure functioning of the system from the start of operation and to protect themselves from any possible risks during installation.

7.11 START UP AND CHECKOUT PROCEDURES

These procedural steps are very important and the installed system is accepted for operation, after these steps are successfully performed.

Step 1: *Check each of the subsystems; the PV array and power conditioning electronics.*

The array checkup involves the measurement of leakage current to ground and open-circuit voltage for each source-circuit. If the leakage current exceeds about 10 micro amps per module, it is an indication of a short to ground in the module and it needs replacement. The measured value for each source circuit must be similar and to ensure identical illumination conditions, Voc is determined for all the source-circuits in quick succession at noon on a clear sunny day. Another preferred method of measuring Voc is by studying the current-voltage curves. The power conditioning electronics,- charge controller and inverter, are checked following the test procedures supplied by the manufacturer.

Step 2: *Checking of all System Control Functions- In order to ensure proper system function and safety, the system operating control functions and array electrical output are tested. The electrical energy output measured immediately after installation would be an indicator for the service personnel in the future maintenance.*

The control functions to be verified are 'normal startup sequence', 'manual shutdown', 'automatic morning startup' ,'automatic evening shutdown', 'shutdown with loss of utility', 'startup following loss of utility', 'manual utility disconnect', 'manual array disconnect' and other performance indicators .Some of the functions are relevant to grid-connected PV systems.

The electrical energy performance of the system under specified conditions of temperature and solar insolation is tested by measuring peak power tracking, the efficiency of the power conditioning electronics and the system power output design rating.

Since the system temperature, the solar insolation and wind level generally do not match the 'standard conditions' the measured energy output may not be same as the designed power. However, if all the control functions are normal during tests, then a mismatch of about 10% between the measured output and the design output if measured on a clear sunny day at noon, is taken as normal functioning of the system.

7.12 OPERATION AND MAINTENANCE GUIDELINES

The system owner and the service personnel have to follow specified guidelines to keep the system under perfect operation during its lifespan, though a PV system normally requires very little maintenance.

The owner should have a broad understanding of the system and the ain components. He/she should ensure that

(i) *no portion of the array is shaded by any obstacle,*

(ii) *the dust settled on the array be regularly removed,*

(iii) *nobody walks on the array during cleaning,*

(iv) *no meddling with electrical parts if not familiar and*

(v) *follow the owner's manual in times of malfunction or emergency. The electronics and the batteries should be kept clean, dry and free of dust.*

The maintenance personnel should be familiar with the safety measures and the trouble-shooting charts. Regular maintenance and inspection should be carried out by measuring system output and the power conditioning system. Module failure is the most common cause for the system which happens due to cracked cells or failed solder joints or interconnects not-properly soldered to cells or cell string shorting to frame or broken interconnects. Measurement of voc and Isc would help to recognize the problems of an array. The personnel must be trained in different techniques to locate the faulty module and to replace it which requires a knowledge of the design specifications.

Battery bank is another component which requires constant monitoring. The manufacturer's instructions must be strictly followed. Accumulation of gases on top of the batteries should be avoided. Corrosion of battery terminals must be checked and corrected. Charging and discharging must be properly carried out; and the voltages and currents should be periodically checked to know the status of the batteries. These steps assure long life for the batteries.

The installation, operation and regular maintenance of a PV system are as important as the designing of the system for successful delivering of electric power over its designed lifetime.

REFERENCES

1. ***A.Mani and S.Rangarajan:*** *The solar radiation over India, Allied Publs., New Delhi, 1982.*

   a. ***www.solarpathfinder.com/doc How_to_doc_solar_radiation_nasa_data.doc***

   b. ***www.pv.unsw.edu.au/research/idl.asp***

   c. ***University of Massachussets - Lowell - Intl. Solar Irradiation Data Base***

2. ***The Design of Residential PV Systems,*** *Vol.4, Sandia National Labs., USA, 1988*

3. ***The Design of Residential PV Systems,*** *Vol,2, Sandia National Labs., USA, 1988.*

CHAPTER EIGHT

BUILDING INTEGRATED PHOTOVOLTAICS

8

BUILDING INTEGRATED PHOTOVOLTAICS

8.1 INTRODUCTION

Solar photovoltaics is one of the few options to produce electricity with no emission of harmful gases and noise in urban areas. Cities have unique and significant potential to exploit PV electricity because they not only have a focussed and large energy demand, but also a physical infrastructure that can support localized power generation. .Building integrated photovoltaics(BIPV) is an energy concept started in 1980 in which PV modules form an integral part of the building. BIPV can be considered as a building element and as an electric power generator from sunlight. The versatility and modularity of PV systems make them easy to integrate into the building .Since aesthetics are an important aspect of a building from the architectural point, many attempts have been made in the last one decade to improve the aesthetics of PV modules. The conventional building elements such as roof tiles, cladding materials, facade elements, shading elements, window pane can now be replaced by appropriate PV modules, and can enhance the overall appearance of the building.

BIPV is expected to be the largest application for grid-connected Photovoltaics since it can be installed in the existing buildings and also be integrated into new structures. (Ref.1). BIPV is considered to be the most promising PV application to bridge the way from small applications to major power generating source of the present century (Ref.2).

8.2 BIPV COMPONENTS AND SYSTEMS

BIPV systems are installed as stand-alone or grid-connected systems,- more often they are grid-connected. The types of systems under utilization are (i) sloping roof systems, (ii) flat roof systems where the modules are placed in concrete or plastic box, (iii) facade systems wherein the modules replace large glass surfaces, (iv) integrated systems as facade accessories in which the modules are arranged as shading or solar protection systems, (v) urban furniture such as bus shelters, public parking structures and (vi) isolated structures which are suitable to integrate PV modules.

The PV roof tiles or shringles are elements with integrated solar cells used for roof covering, together with standard roof tiles or shringles. They are tightly integrated into the roof. Many different forms are used: PV roof tiles, PV roof shringles, solar laminates, modules with integrated solar cells produced as roof covering elements, transparent laminates or modules for flat roofs etc.

PV roof tiles with mano-or poly Si cells are used with the clasical roof tiles of the same producer and PV roof shringles with crystalline/ thin film solar cells. They are characterized by high efficiency due to ventilation on the rear side. Some support structures used for conventional roof tiles can be utilized for PV roof tiles also. In the case of PV facades in vertical buildings, the energy gains are reduced due to non-optical orientation and differing tilt angles (Ref.5). The table shows the energy gains at different tilt angles and orientation of the PV facade.

Table 8.1

	Energy Gain (%)		
Orientation of PV facade	**Tilt Angle: 0°**	**30°**	**90°**
West-East	100	98	66
Southwest-Southeast	100	106	73
South	100	108	73

Several manufacturers are producing PV modules which can substitute the expensive traditional building components; and these modules are able to aesthetically integrate into the building and decrease the cost of construction and the power distribution losses. Enhancing local power quality, pleasing visual appearance and potential utilization of the building grid are the other advantages that these modules offer.

The development of coloured PV cells is an added advantage to BIPV. Colours such as green, magenta, gold pink and steel blue are available with different thicknesses of antireflection coatings. Semi-transparent PV cells and modules are now available which help to provide homogeneous day lighting of rooms. Highly efficient semi transparent solar cells with transmittances ranging from 10 to 30 percent averaged over the whole cell area are available. Semi transparent a-Si solar cells are also manufactured now. These semi-transparent cells are highly suited for shading elements, facades and roof windows. PV modules are also available to replace windows and glazing elements in warm facades and roof elements.

Several types of glass panes are produced: thinnest possible pane of low-iron content and highly transparent glass to allow solar irradiation with almost no loss are available for front glass panes. For the back glass pane, the options are plenty,- solid coloured glass, colour-coated glass depending on the customer's choice, silk-screen printed glass where the pattern and colour can be chosen, coated glass where metal oxide layers are coated on the glass backside which have a reflective effect, insulation glasses that provide heat insulation to the building, noise-protection glasses and sun protection glasses with a selectively reflecting metal-oxide layer on the backside that reflects the heat energy in the solar radiation and allows visible light only. Special laminates with coloured back sides have also been produced . Due to sefety requirements for lamination, usually PVB foil instead of EVA foil is used.

Laminate can consist of mano-Si cells, thin film cells or transparent cells.

Customised modules include module shapes, cell type and colour, cell transparency, laminate construction, laminate/module size, heat/noise isolation properties, module voltage and peak power.

Significant improvements have been achieved in BIPV with regard to PV module quality, PV roof tiles, flat roof integration, PV in facades, colored PV cells and semi-transparent modules.

8.3 SYSTEM DESIGN

Various roof integrated PV system designs have been developed and installed. A typical design developed by an Australian team (Ref.3) consists of a metal sub-tray water proof roof sheet on which high efficiency frameless PV module laminates are fastened using custom-designed diamond-shaped mounting clips. A special design element allows natural convection cooling between laminates.

The modules are surrounded by a mesh to prevent small animals or other unwanted materials entering behind the modules. Weather proof between trays is also ensured. BP Solar module laminates are used; and to satisfy aesthetic demand, a dark coloured backing layer is chosen resulting in the reduction of efficiency from STC by two percent. The design is such that the tray installation and PV wiring could be completed in about two and a half hours.

The merits of the design are that (i) each system delivers 1600 kWh per year, (ii) it complies with the Australian building load, health and safety specifications and electrical standards and (iii) it provides a simple design for rapid installation of PV systems causing no obstruction/delay in the construction of each building. The frameless PV laminates and diamond-tray mounting clips are considered to lower life-cycle energy costs and ensures a balance of cost, thermal performance and energy-yield. This design is simple, easy to install and meets the market requirements.

8.4 FUTURE PROSPECTS

BIPV demonstrate immense potential in the urban areas due to restricted land availability. In addition, the building surface area available is very large. In developed countries, there is a great deal of interest in BIPV applications of various types. The biggest potential application for PV in Europe is integrated energy generators installed in the built-environment and connected to the load grid. The market potential in Europe is growing at 33% per year (Ref.6). Globally BIPV are expected to account for half of the total PV market share by the year 2010, and in Europe the percentage is even higher (Ref.7). The avoidance of land use, development of advanced construction materials and integration of PV into the total costs of the building are the contributions from PV technology to the new urban building constructions.

The close association and working of PV industry with architects and building industry should result in the development of cladding elements, prefabricated building integration arrays, facades made of multi-functional modules with air circulation and thermal and aesthetic qualities, multipurpose building elements of glass-glass layers with cells of different shapes to suit architectural designs (Ref.3) in the coming years.

The following product lines need development and improvement in fabrication and application:

(a) *flexible, light-weight roofing materials,*

(b) *peak-shaving systems,*

(c) *modular homes manufactured with PV,*

(d) *modules that generate AC and*

(e) *architectural PV glazing systems.*

The outcome of these developments will significantly change the architectural features and constructional methods of the buildings, and provide support to the concept of localized generation of non-polluting power and supply with negligible distribution losses.

The other supporting activities under study which will further enhance the application of BIPV are:

(i) *educating architects about the emerging PV applications in buildings,*

(ii) *assessing the benefits of PV as a demand-side management option,*

(iii) *studying solar resource/utility load matching,*

(iv) *investigating PV power generation matching the individual building loads.*

8.5 TYPICAL CASES

Thousands of buildings, existing and new ones, around the world have incorporated BIPV Systems into them to meet part or total power requirements of the buildings. These Systems not only provide quality power but are environment friendly. Many BIPV demonstration projects are installed around the world, a large number being in Europe. To cite a few: (i) Solarex Corporation PV module manufacturing plant in USA is partly powered with their own PV modules integrated into the plant's roof. The System installed in 1982 consists of PV array of 200 kWp capacity and a battery bank to provide uninterrupted power supply, (ii) The Munich Trade Fair Building has part of its huge roof area covered with mono-crystalline silicon frameless modules of capacity, 1016 kWp. This system , the world's largest at the time of installation, is mounted on the ribs of the roof and has been working since its commission in 1998,(iii) In Solar Olympic Village built for Sydney Olympic Games in 2000, 629 homes are equipped with integrated 1 kWp PV systems on the rooftops to meet the power needs, (Ref.3). Other PV projects under this PV initiative are installation of nineteen PV lighting pylons and a 70.5 kWp PV installation using a-Si panels on the Superdome above the banqueting hall. The main objective in the use of solar power in the Olympic village has been to develop the world's largest solar-powered suburb offering an eco-friendly model for future urban development.

Other structures that are suitable for installing PV systems are motorway and railway sound barriers, bus shelters and filling station canopies. The demonstration systems for sound absorbing walls on motorways have been installed in Switzerland (100 kWp capacity),in Austria (40 kWp capacity), and in Netherlands (55 kWp capacity).

It is noticed from a Swiss survey that suitable sites just on the side of motorways and railways for installing PV would give a capacity of 375 MWp which could be a very significant power source in the country (Ref.4a). The progress in BIPV has been very significant in European countries and in North America.

Increased feed-in tariffs and/or special grants in recent times in many countries – Spain, South Korea, Italy, France, Switzerland, Japan - are making it more attractive to utilize PV building elements that generate electricity while fulfilling at least one more function in the building. In Japan, more than 10 MW of PV modules have been integrated into roof tops in various forms – from high, narrow and elongated modules via rectangular standard module forms up to square solar roof tiles (Ref. 4b).

Some typical examples of these installations are briefly mentioned..

(a) Renovated building of Ministry of Economic Affairs, Germany:

This is an existing structure of 18th century not in use for long in Berlin.The renovation has been taken up after the unification of two Germanys. Not only historical and visual aspects, but ecological aspects also have influenced the re-creation of the building. As a part of refurbishing, PV systems are included in the renovating plan.

A huge PV installation installed on the inclined roof covers a 180 meter-long inclined façade of area about 900 sq.meters. The System of capacity 100 kWp is made from insulating glass. Each of the 1.4x2.7 meter PV elements consists of two 6 mm panes from the glass-glass module and 12 mm laminated glass on the inner side.

Some part of the façade is transparent while the entire upper half comprises semi-transparent PV modules. The construction provides solar- and noise-protection , shading and solar electric power of about 85,000 kWh per year. This system was installed in 1999 and is a very good example of a successful combination of the use of solar energy and a typical renovated building (Ref. 8).

This building is a part of a cluster of buildings to house the Parliament and offices which also have either rooftops or building integrated photovoltaics. They include Chancellery with 150 kW, Presidential Administration building with 44 kW and the Parliament building with 36 kW on the roof. These new and renovated buildings are located around the Parliament building

(Reichstag building) and are connected to a local renewable energy electric grid that includes about 400 kW of rooftop and BIPV.

(b) Coloured PV Façade in Prague, Czech Republic:

One of the first installations of coloured PV modules in the world has been erected on a hotel building in the city of Prague in April 2000. A 6 kWp golden colour PV array has been installed on the south-facing façade of the building. The Array comprising 132 modules covers an area of 66 sq.meters. The modules are fabricated with mono-crystalline silicon cells with gold colour on the active side and are rated 45 Wp. Three grid-connected inverters each of nominal power, 1500 W are used in the System. The System is expected to deliver an electrical output of 4.6 MWh annually with a lifetime exceeding 20 years. A monitoring and visualization system is also installed for continuous data monitoring and real-time data visualization (Ref.9).

(c) SMUD's PV Pioneer Programme

The Sacramento Municipal Utility District in California, USA, has worked out a unique programme called PV Pioneer 'green pricing' programme, where in SMUD would partner with the householders in installing PV systems.

The partnership is so designed that SMUD purchases, installs, owns and operates residential rooftop PV systems.(Ref.10). This programme has been aimed to promote PV commercialization as well as to develop roof tops as PV power distributed generation sites.

This is one type of BIPV developed for residences and small commercial buildings. 436 residential systems each of 3-4 kWp capacity have been installed during1993-96, amounting to a toal capacity of 1.9 MWp. This programme has continued and around 178 kWp in 2000 and nearly 1200 kWp in 2001 has been installed on more than 220 residential homes and commercial complexes (Ref.11).

A typical 4kW system uses 400 sq. ft. PV array which is fixed to the roof using innovative roof jacks (Fig. 8.1) The components are so designed that the system installation could be completed in half a day only. This programme has become very popular in California.

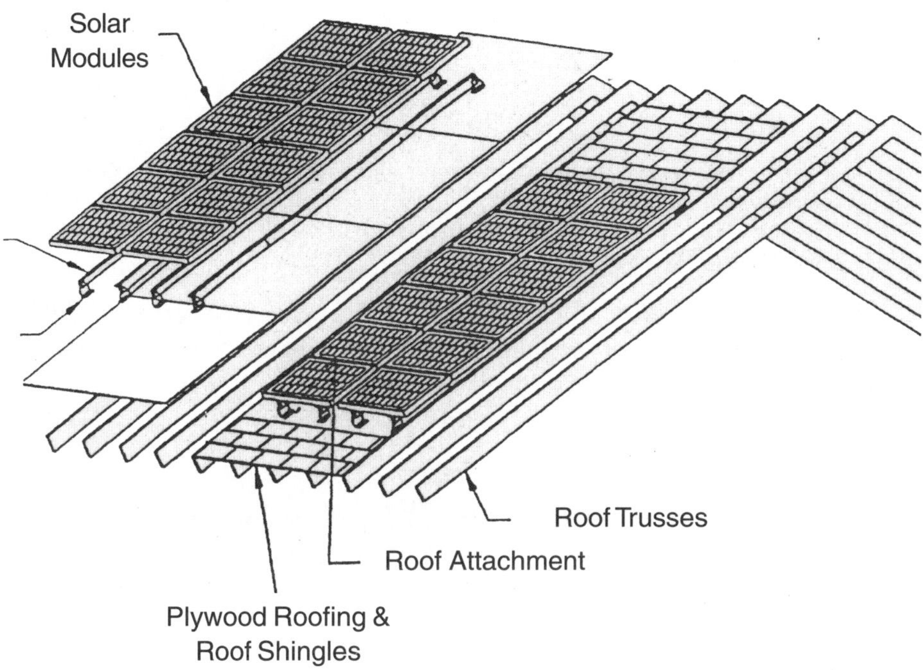

Fig. 8.1 *Conceptual PV/Roof Interface (Source: SMUD User's Leaflet)*

(d) Largest Solar Roof in Amsterdam, Netherlands

The World's largest solar roof, 2.3 MWp capacity has been installed on the Floriade Exhibition Building in Amsterdam in October 2002. This system comprises 19000 transparent solar panels covering an area of 26100 sq.meters. and connected to the grid. It is expected to generate 1.23 GWh per year. The building holds horticulture exhibition; the plants and flowers thrive under 26 nos. of 100 sq.meter of PV panels which are specially designed for the light to filter through. (Ref.6).

(e) Solar Car port in California, USA

The largest solar car roof has been installed in Riverside, California in 2001. The PV array consists of 2016 modules each of 75 Wp capacity. The system, connected to the utility grid, is rated to provide 113 kW AC output under typical conditions, approximately enough to provide power to 57 homes. The roof provides shade for approximately 152 parking spaces at the Operation center for Riverside public utilities.

(f) Recent BIPV Projects

The following three recently completed BIPV projects are given as examples which are not only successful combination of integrated PV and aesthetically attractive architecture, but perform additional functions such as sound insulation, heat insulation and shading.

(i) a solar greenhouse in 'Southeast Germany'

(ii) a suburban railway station in Enschede, Netherlands which is redesigned and equipped with semi-transparent, curved solar roof with crystalline solar cells,

(iii) a semi-transparent PV façade with bifacial cells and solar sliding shutters in a private house, performing multifunctions.

The details are given in Ref. 4b.

8.6 CASE STUDIES

The Northumberland Building PV façade installed in 1994 as one of the demonstration projects has been one of the many cases which have worked successfully from the date of installation. The status report of the performance over the first 5-year period,1994-99 is briefly outlined (Ref.12). The main advantage in integrating PV systems into a building is to offset some of the capital cost by replacing the traditional building materials by PV modules; and this advantage can be realized if the system successfully performs over the productive lifetime of more than 25 years. An extensive monitoring would help to understand the performance and such studies would naturally promote confidence among the architects and engineers of the building industry to implement BIPV systems. (Plate 1)

In the design of the façade, the PV modules are fully integrated into the rain screen overcladding on the south façade of the building, which is tilted at 25 degrees to the vertical.

A conventional electric circuit design is used with 15 modules connected in each series string and 31 strings connected in parallel. The nominal array rating is 39.5 kWp. The output from the system is fed into a 40 kWp inverter at a dc input voltage of about 250 volts. The 3-phase ac output is fed into the building power supply system at 415 volts. The system is affected by shading due to the surrounding buildings, reducing the electrical output by a varying amount depending on the time of day and season.

During the 5-years of operation from the day of installation in late 1994, the system has provided over 100 thousand kWh to the building. Extensive monitoring of the system has been done and the data analysed to provide information on

(i) *annual and monthly outputs,*
(ii) *climate variations,*
(iii) *long term perform changes,*
(iv) *maintenance procedures and*
(v) *component reliability*

(i) *Typical System Outputs*

The system output has been measured month-wise for the five year period. The outputs correlate well with the measured in-plane insolation levels. The performance has been found to be good during the spring and autumn months. However, the March data indicates that there is considerable variation from year to year due to variable weather conditions. A significant reduction in the output is observed in the winter months because of increased shading.

(ii) *Climatic Conditions*

At the design stage, the historical solar data is used and it is seen that the system is mainly influenced by this data. The measured data during the 5 years is compared to see how closely they match. The measured insolation data is compared with the monthly and hourly values provided from Meteonorm for Newcastle where the system is located. The Meteonorm monthly data appears to overestimate the summer sunlight levels. But underestimate the early spring (February) and autumn (October) levels. This difference is most likely due to the assumptions regarding the balance of direct and diffuse sunlight in the translation of data to the inclined surface. It is also noticed that the hourly analysis than the monthly analysis from Meteo-norm matches better with the measured data, even when only monthly values are considered.

The operating temperatures are measured for each month and compared with the average monthly ambient temperatures; and they are found to relate linearly. This observation helps to predict the average operating temperature for the System from the knowledge of average ambient temperature which will be useful to predict the system performance. The maximum operating temperature recorded for the system is 60°C.

(iii) Long term Performance

The monthly performance efficiency over the five years of study has shown considerable variations from year to year, specially in autumn and spring months, though a clear trend is not seen. The table indicates the measured output along with the predicted output calculated on the the basis of first year performance and correcting for the measured sunlight levels and inverter efficiency. The differences for all the years is within 5%. This variation is attributed to the influence of shading and light levels. The Average system efficiency over the period is 8.1% and the average Performance ratio is 0.59. However, if the correction is made for shading effect the efficiency is expected to be around 10 to 10.5%.

Table 8.2

	1995	**1996**	**1997**	**1998**	**1999**
Measured Output (kWh.AC)	21,836	17,460	21,435	18,834	20,730
Ave daily in-plane Insolation(kWh/m^2)	2.6	2.1	2.4	2.3	2.5
Annual System Efficiency (%)	8.2	8.0	8.6	8.0	7.9
Ave Inverter Efficiency (%)	90	88	90	90	90
Predicted output (kWh.AC)	–	17,516	20,488	19,128	21,413
% difference from the measured	–	+0.3	–4.4	+1.6	+3.3

Flat Roof Integrated PV Installation on the Convention Centre, Basal, Switzerland (240 kW_p)

The world's largest solar roof, at the Floriade exhibition, Amsterdam. Photo: ANP/Nuon

Anne - Sophie School (insulation glass, semi-transparent, CIS)
(By Courtasy of B. Dimmler, Weurth-Solar)

A Residential building with roof integrated CIS modules
(By courtesy of B.Dimmler, Weurth-Solar)

Beijing Railway Station (the world-wide biggest building-integrated CIS installation
(By courtesy of B.Dimmler, Weurth-Solar)

Plate 1 *PV Facade at the Northumberland Building*

(iv) Reliability

The PV façade has exhibited very high hardware reliability over the operating period to date. The replacement of two bypass diodes in the first summer and a trivial problem with the inverter are the only problems observed. The system has been down for less than 48 hrs during the entire operating time. The system is cleaned annually; it is felt that the dust accumulation is not a significant factor to affect the system efficiency and hence the annual cleaning has been considered adequate enough.

In conclusion, the performance study of the first five-year operating period is helpful to validate the use of this information for the lifetime of the system, and to establish the likely variations which may take place during that period. The analysis, however, is highly encouraging from the point of consistency of performance and reliability of the system.

REFERENCES

1. ***R. Hill:*** *Intl. Conf. Photovoltaics, Clean Energy Today, ETSU(London), (1996)*
2. ***F.Sick et al:*** *Photovoltaics in Buildings, A Handbook for Architects and Engineers, Intl. Energy Agency, Paris, (1996)*
3. ***Deo K. Prasad:*** *Renewable Energy World, 3, no.4, p77(2000)*

4a. ***M.Oliver et al:****Energy Policy, 27, p371(1999)*

b. ***Solar and Wind Energy:*** *BIPV-Boundless possibilities,p94(1/2009)*

5. ***Andreas Hanel:*** *Renewable Energy World, 3, p89(2000)*
6. ***Erik Lysen:*** *Renewable Energy World, 6, no.1, p43(2003)*
7. ***Keon Steemers:*** *Renewable Energy World, vol.4, p106(2001)*
8. ***Eric Schaar-Gabriel:*** *Renewable Energy World, 2, no.6, p39(1999)*
9. ***News item:*** *Renewable Energy World, 3, no.3, p130(2000)*
10. ***D.E.Osborn:*** *American Solar Energy Society, Solar 97, April(1997)*
11. ***News item:*** *Renewable Energy World, 5, no.2, p22(2002)*
12. ***N.M.Pearsel et al:*** *16th European PV Solar Energy Conf., Glasgow, p1804(2000)*

CHAPTER NINE

ENERGY ANALYSIS AND ENVIRONMENTAL ISSUES

9

ENERGY ANALYSIS AND ENVIRONMENTAL ISSUES

9.1 ENERGY ANALYSIS

9.1.1 Introduction

Energy analysis is an important study developed in early 1960s to determine the energy flows in industrial processes; it amounts to the calculation of the required energy input for unit output of the product. This concept has been used to assess various PV cell technologies and production methods in the last three decades.

A production technology consists of several process steps and the energy requirement for each step is calculated, which are then summed up to give the energy requirement for the concerned production technology (Ref. 1).

In each process step, three quantities, (i) materials, (ii) machinery and (iii) processing are involved. The first one is the amount of energy consumed in the prepartion of materials that are present in the finished product. The second one is the amount of energy consumed in the manufacture of equipment used in the production. This component is estimated by relating the energy content directly to the capital costs of the equipment. The third one is the amount of energy used to perform the process step which is mostly in the form of electrical energy. However, materials such as gases used in the process are also included, though they do not appear in the final product. From the total energy requirement, 'the energy pay-back time' and 'the energy ratio' of a production technology are estimated.

In a PV technology, the energy pay-back time is the time taken for a particular type of PV module to pay back the energy required in its manufacture. This depends on the solar insolation level which in turn depends on the location of operation of the module. The energy ratio is the ratio of the total amount of energy produced during the lifetime of the module to the amount of energy consumed in its production which is dependent on the location of operation of the module and its expected lifetime. A lifetime of 25 years, consistent with the expectations of the module manufacturers, has been assumed in most of the studies. The energy analysis results are expressed as Kwh/kg of silicon or MJ/cell.

9.1.2 Energy Analysis of PV Cells and Modules

Several research groups studied the energy requirement of mono- and multi-crystalline silicon cells. Hunt (Ref. 1) has estimated the energy requirement at different processing stages from raw materials to finished product for silicon solar cells and has arrived at 112.3 MJ/cell for the energy requirement, and 12 years for energy payback time for terrestrial silicon cells. The PV industry has accepted his results considering them as reliable. Fischer (Ref. 2) has studied the energy requirements for the production of poly (multi-) silicon cells by the three main methods used at that time and has concluded that 'casting the ingot' and then producing wafers are the least energy intensive processes with an energy requirement of 3.9 MJ/cell or 2290 Kwh/kg. Wolf (Ref. 3) has calculated the energy requirement for silicon as 2560 kwh/kg. Iles (Ref. 4) has studied the energy requirement from Czochralski growth to the testing of silicon solar cell. Riddoch et al (Ref. 5) has worked out the energy requirement for amorphous silicon as 3.6 MJ/cell for an MIS cell with a-silicon deposited by RF glow discharge of silane.

Hagedorn's work (Ref. 6) on hidden energy in silicon solar cells/modules and power stations has been the basis for energy studies of silicon for about a decade since 1989. The energy payback times for mono- and multi-crystalline silicon and amorphous silicon calculated for two scenarios vary between 1.7 to 6.0 years for cells/modules, and 7.3 to 2.6 years for power stations. Palz and Zibetta (Ref. 7) have extended these studies and concluded that energy payback times of this order could be reduced by further industrialization of the processing technologies. The current payback times are comparable to those of electricity generation in fossil fuel and nuclear power plants.

Early studies of Jenkins et al (Ref. 8) and Bloss et al (Ref. 9) on thin film technologies have concentrated on CdS/Cu_2S solar cells. The former have estimated the energy requirement for sputtered CdS cells as 1.1 MJ/cell, while the latter as 1.4 MJ/cell for thermally evaporated CdS layer. Hynes et al (Ref. 10) have studied CIS-based cells produced by thermal co-evaporation and electro-deposition technologies and their energy requirement values are 400 kwh(thermal)/m^2 for thermal co-evaporated cell and 250 kwh(thermal)/m^2 for the electrodeposited cell. This study includes an examination of how the energy requirement is sensitive to the various process parameters as well as the effective CO2 emissions for these two technologies. Hynes et al have also analysed (Ref. 11) the energy requirement of thin film CdTe modules processed in two types of technologies – non-vacuum processing and vacuum processing. The energy requirement for the CdTe modules based on the first process is 275.7 kwh(thermal)/m^2 and 329.9 kwh(thermal)/m^2 for the second one. These values lead to energy payback times of 0.6 years for a 10% module operating in southern Europe, and an 'energy ratio' of 24 for non-vacuum processed modules; and for vacuum processed modules, 0.8 years (energy payback time) and 20 (energy ratio).

Alsema and co-workers have published energy analysis studies since 1990s on several technologies – crystalline and amorphous silicon, CdTe and CIS thin film cells(Ref. 12, 13). Their studies show an energy requirement of 399 kwh(thermal)/m^2 and energy payback time of 1.3 years for multi-crystalline silicon, and 224kwh(thermal)/m^2 and 1.2 years for amorphous silicon. For thin film materials, 304 kwh(thermal)/m^2 and 1.1 years for energy requirement and energy payback time respectively. For single-crystalline silicon, this group (Ref. 14) has found 47 to109 MJ/cell depending on the method of feedstock production. The energy requirements for balance of system components, and the energy payback times for different types of modules used in different applications for the current and the expected future technologies are also discussed in this paper. In their recent energy analysis studies carried out on ribbon, multi and single crystalline silicon from the reliable and accurate data obtained from several PV manufacturers, they have reported 1.5 to 2.5 years for the energy payback times for southern European locations (Ref. 15, 16).With improvements in the technology, reductions in energy requirements up to 25% are expected, which will further reduce the energy payback times.

The investigations of Knapp and Jester (Ref. 17a) have resulted in energy payback times of just 3.3 years for single-crystalline silicon cells, which includes the energy to make Al frame and the energy to purify and crystallise silicon, and just under 2 years for CIS thin film cells, and energy ratios ranging from 9 to 17.

9.1.3 Energy Analysis of PV Systems

Alsema calculated the payback time without including the energy used in crystallizing microelectronic scrap; his best estimates of electricity used to make frameless PV were 600 kwh/sq.m for single crystal silicon modules and 420 kwh/sq.m for multi crystalline silicon. Assuming 1700 kwh/sq.m per year of available solar energy and 12% conversion efficiency, he calculated a payback time of about 4 years for 2004-year multicrystalline silicon PV systems (Ref.18a). 10 years from now, he assumes a solar-grade silicon feedstock and an efficiency of 14%, and finds payback coming down to 2 years.

Peharz and Dimroth (Ref. 18) have investigated in detail the individual components and their 'cumulative energy demand(CED)' of a high concentrated PV system. The energy payback time has been evaluated as 8 to 10 months for a system operated in Spain, and 12 to 16 months for a system installed in Germany. For CdTe modules (2004 technology), with the data from German manufacturer, ANTEC, an energy pay back time of 0.9 years has been estimated (Ref. 19).

Hynes et al (Ref. 20) investigated the energy analysis for the system of capacity 39.5 kWp installed on the Northumberland building in UK(which is discussed under BIPV in this book). The total energy requirement of the system was 561102 kwh(thermal), giving an energy payback time of about 4 years and energy ratio of about 7. This study shows that for a northern location and for1994 PV technology, it is possible to have a low energy pay back time for a demonstration system.

A residential PV system of 3 kW studied the crystalline, polycrystalline and amorphous silicon modules (Ref. 21a). Energy payback times of 3.3 to 11 years for mono-cyrstalline, 1.5 to 2.4 years for the polycrystalline and 1.1. to 2.1 years for the amorphous silicon were obtained.

The notion that ground-mounting support structures, such as concrete foundations and heavy framing would increase the payback times is incorrect according to Mason (Ref.21 b). By not including the modules, he has found that the Tucson Electric Power's Springville's multi crystalline silicon power plant has paid back the energy invested in the ground-mounted PV system in about 3 to 5 months. This is essentially same as for a roof-mounted PV system.

Energy payback estimates for both roof top and ground mounted PV systems are approximately the same, depending on the technology and type of framing used.

The energy payback times for different cell technologies at different locations are shown in Fig. 9.1 (Ref.21c).

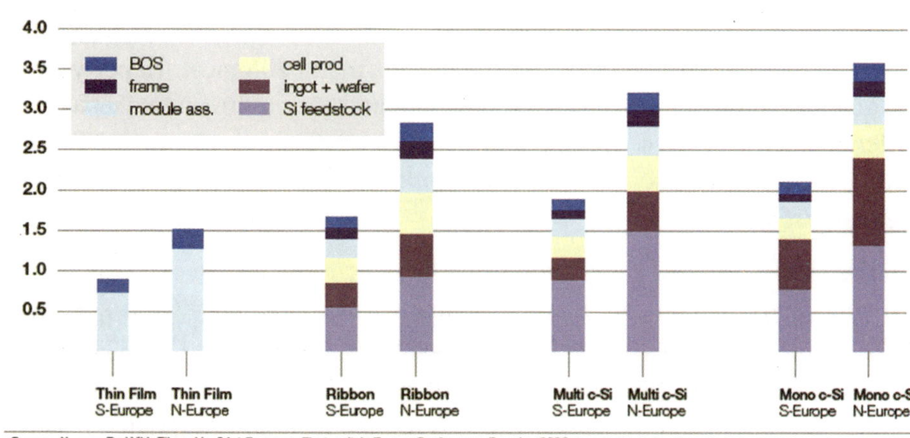

Fig. 9.1 *Energy payback times for range of PV systems*

It is interesting to see from the studies of Dones and Frischknecht (Ref.21d) that the PV systems fabrication and fossil-fuel energy generation have similar payback times (including costs for mining, transportation, refining and construction.

9.2 ENVIRONMENTAL ISSUES

PV manufacturing volumes are moving from mega watt to giga watt level. At that level, it becomes imperative for the industry to look at the environmental issues associated with PV manufacturing processes and find solutions to ensure that PV continues to be a sustainable industry.

9.2.1 External Costs & Life Cycle Analysis

The PV solar technology is promoted as sustainable energy technology; however, it suffers from a disadvantage of high energy consumption. It means, PV power generating systems are not fully capable of mitigating greenhouse gas emissions (Ref. 22a).

The ExternE project funded by European Commission defines the 'external costs' as follows: It is 'the monetary quantification of the socio-environmental damage, expressed in eurocents/kwh, with the possibility of providing a scientific basis for policy decisions and legislative proposals such as subsidizing cleaner technologies and energy taxes to internalize the external costs.' (Ref. 1). The reader may refer to http://europa.eu.int/comm/research/energy/pdf/externe_en.pdf. for full details of the project. Baumann et al (Ref. 25a) describe basic assumptions and the requirements of the methodological framework for the quantification of external costs, and compares the environmental effects of different energy technologies including the conventional (coal, oil, nuclear and natural gas) and the renewable. The environmental costs of energy technologies are discussed in detail.

The recent ExternE study shows that PV electricity has an external cost of 0.6 cents/kwh, as compared to 1-2 cents/kwh for electricity from gas (Ref. 25b). These estimates are however unacceptable because the PV data used is not representative of the current PV technology (Ref.25c).

The environmental impacts of PV technology were not seriously taken till 1980s. For the first time, at a German-American workshop, 'External Environmental Costs of Electric Power: Analysis and Internalisation' held in October 1990, the researchers considered environmental damage and ways in which environmental costs may be internalized for wind energy, which could be relevant for the PV technology.

An extensive review of environmental costs/risks covering all electricity generation technologies including PV installations was published by Ottinger et al (Ref. 22b). Several environmental costing models were also discussed and based on this, many studies on external/environmental costs have followed. Sorenson's extensive report (Ref. 23) on life cycle analysis outlines the principles and methodology for life cycle assessment of renewable energy technologies and applied to multi crystalline and amorphous silicon cells (Ref. 24).

Life-Cycle Analysis: Life-Cycle Assessment (LCA) is an analytic methodology generally employed to make environmental decisions in the manufacturing sector. It is a comprehensive inventory of inputs and outputs along with the life of a product (from raw material extraction and energy use through ultimate disposal of residues), combined with an analysis of identified impact and potential for their reduction (Ref. 26 a). Several hundreds of studies are published on this subject. A few important ones are briefly mentioned here. Alsema et al (Ref.26 b) have studied CdTe and CIS thin film modules by means of life cycle assessment of Cd, Te, Se and In material flows. The energy analysis of these module types and emissions of Cd and Se are also investigated. These emissions are found to be of the same order of magnitude as those from a modern coal-based power plant. van Engelenberg and Alsema (Ref.27a) have also studied the environmental risks of amorphous silicon modules. Their life-cycle assessment using material and energy flows has not shown any serious risks coming from the production and use of amorphous silicon modules. Concerning the future status of PV technology, a worst, base and best cases are considered for analysis. The results show that significant releases of tin oxide, chloride and fluoride are not specific to these modules, and risks concerning silane storage and handling require serious attention. Alsema et al (Refs. 27 b, c) have also investigated an environmental life cycle assessment on 'advanced' production technologies for multi-crystalline silicon modules along with 'standard' technology (i.e. silicon from the Siemens production process, standard casting and sawing, cell processing, module assembly in a plant of 10-30 MW/ year capacity). The two 'advanced' technologies utilize solar-grade silicon feedstock produced by Solsilc and Bayer/Deutsche solar processes, and wafers of 300 ìm thick are prepared with Ribbon Growth-on-Substrate (RGS) or Edge-defined Film-fed Growth (EFG) methods. The cell fabrication and module assembly are done in a plant of higher capacity, 30-100 MW/year, enabling a saving of about 10% in water and energy consumption. Several alternative BOS concepts and inverter types are also tried. The results show that the two 'advanced' technologies score a 50% reduction in the environmental impacts of the modules compared to modules of the 'standard' technology. About 60% reduction in greenhouse gas emissions is possible with advanced technologies, and the Systems are significantly better than the ones with standard technology (Ref.27c).

Hynes et al (Ref.28) have studied life-cycle analysis of PV modules based on CdTe. Their results show that the production of modules and the waste products disposal are the potential threats to environmental and health factors. But the health and safety regulations can monitor and control or reduce these impacts. With large scale productions, 10-50 MW/year, the supply of tellurium may cause problems due to its limited reserves. This analysis further shows that there are no risks/hazards that cannot be overcome by following standard industrial practices in the production of CdTe modules.

Phylipsen and Alsema (Ref.29) have studied the life cycle assessment of multi crystalline silicon solar cells, concentrating on energy and material flows. The study suggests that:

(a) possibility of recycling silicon wafers both from the production waste and at the end of life, and

(b) methods to reduce the emissions of fluoride in water should be investigated. Further, emissions of volatile solvents and alcohols are to be avoided, and silver usage for contacts should be reduced to avoid resource constraints.'

Recently Komoto et al (Ref.26 a) have estimated energy payback time and 'life-cycle CO2 emission rate' for various kinds of PV systems located in Japan. PV cells analysed are single- and multi-crystalline Si, a-Si/μc-Si, thin film Si, and CIS. PV modules analysed are the framed regular modules for each PV cell. In addition to production, they have considered the recycling and disposal also; three cases are evaluated: 1) without recycling, 2) recycling BOS materials and Al frame and 3) promoting PV module recycling. In the case of 'without recycling', the estimated EPT of residential PV system of 4 kW is 1.4 – 3.0 years, and of public/ industrial building PV system of 10 kW is 1.9 – 3.4 years. The corresponding CO2 emission rates for these systems are estimated as 46 – 78 g-CO2/kWh and 62 – 87 g-CO2/kWh respectively. The difference between residential and public/industrial systems is caused by difference of array structure. These studies are applied to ground-mounted large-scale (MW scale and >100 MW) PV systems (Ref. 26 c, d). The EPT, and total CO2 emission and CO2 emission rates are shown in Figs. 9.2 and 9.3 (excluding after operation).

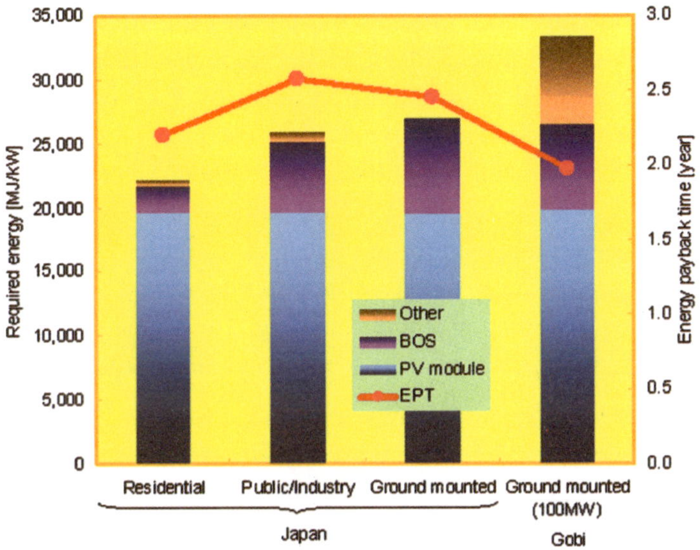

Fig. 9.2 *Required energy and EPT of PV systems, excluding after operation (Ref. 26b)*

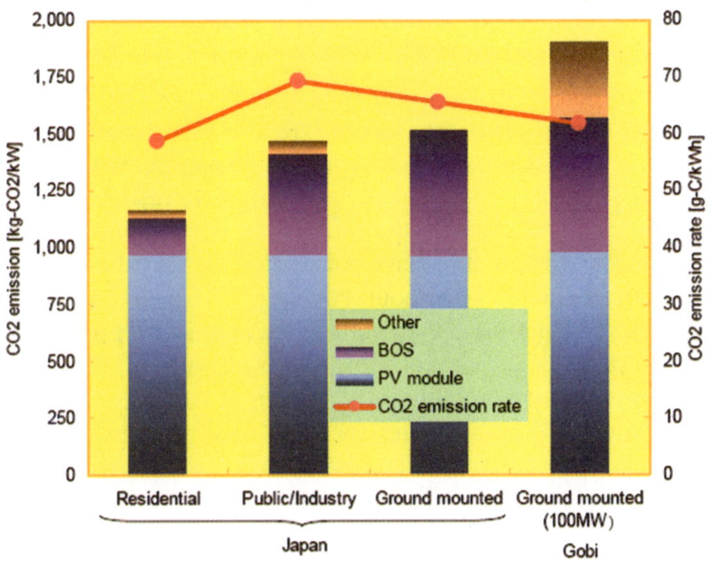

Fig. 9.3 *Total CO2 emission and CO2 emission rate of PV systems, excluding after operation (Ref. 26b)*

Jungbluth et al (Ref. 26e) presents the update and extension of the life cycle assessment of PV power plants in the Ecoinvent data V2.0. Sixteen different grid-connected photovoltaic systems of 3 kWp capacity each, operational in Switzerland in 2005 are studied. The plants differ according to the cell type – single- and multi-crystalline silicon, ribbon silicon, thin film cells with a-Si, CdTe and CIS – and the location. Further, average PV production mix data have been investigated for more than 20 countries. The process data include the full process chain of panel and laminate production, mounting structure, 30 years of operation and dismantling. The region specific EPTs as well as CO2 mitigation potential are evaluated and discussed in detail.

The aim of the European Commission funded ECLIPSE project was to introduce transparency into life cycle analysis and to provide database for recent and emerging PV technologies (Ref. 30) that include single crystal, multi crystalline and amorphous silicon, CIS, and several applications in buildings under different insolation conditions. Complete details of the project are given at http://www.eclipse-eu.

Issues such as environmental aspects of PV power systems, energy payback times, CO2 mitigation potential, life cycle assessment, and health and safety assessment and control were discussed at an Expert workshop in June 1997 in Utrecht. Selected presentations are available (Ref. 31)which include prevention and control of accidental releases of hazardous materials, health, safety and environmental risks from the operation of CdTe and CIS thin film modules, life cycle air emissions from PV systems and life cycle analysis of PV systems in buildings etc.

Outlining the available information databases including ECLIPSE and current studies which may develop the tools required, Fthenakis et al (Ref. 32) argue for an agreed life cycle analysis based on an agreed framework.

9.2.2 Environmental risks/hazards:

Different technologies using different raw materials and production processes are used to produce solar cells. The environmental risks associated with PV manufacturing, therefore depends on the specific type of cell being produced.

The health and environmental issues of concern to PV industry has been continuously published by the researchers at Brookhaven National Laboratory since 1970s, starting with Moskowitz. A comprehensive study of health and environmental issues of the manufacture, use and disposal of thin film cells was published by them in 1995 (Ref. 33). Results of several studies are found in the literature, for example, hazard assessment of the manufacture of CIGS thin film cells (Ref. 34), hazards of CdTe and CIS thin film modules during the operation period including risks due to fire, weather changes or damage (Ref. 35), and hazards of CdTe thin film cells by Fthenakis (Ref. 36) to mention a few.

The Table shows the major hazards in the manufacturing of different types of solar cells (Ref. 37).

Cell type	**Potential risks**
Crystalline Si	HF acid burns, SiH_4 fires/ explosions, Lead solder/ module disposal
Amorphous Si	SiH_4 fires/explosions
CdTe	Cd toxicity, Carcinogenicity, Module disposal
CIS, CGS	H_2Se toxicity, Module disposal
GaAs	AsH_3 toxicity, As carcinogenicity, H_2 flammability, Module disposal

Crystalline silicon: Silicon is relatively harmless and causes almost no danger to the environment or to human health. However, in the production of silicon cells, several chemicals including lead are used which pose a potential risk to environment and to health, if not properly handled. Hydrofluoric acid (HF), nitric acid (HNO_3) and sodium hydroxide (NaOH) which are used in silicon wafer cleaning are caustic. These chemicals are extremely hazardous, but due to their routine use in industry, safety measures to handle them are well established.

The toxic gases such as hydrogen bromide, chlorine, phosphoric acid used as dopants in the production are highly hazardous and affect the health of those within their immediate vicinity. The risks, however, can be minimized by proper use and by maintaining ventilation. Some of the halogenated substances and their byproducts are powerful greenhouse gases, much more potent than the carbon dioxide. Other substances of concern are the lead based solders, cleaning agents and solvents used in cutting, etching and polishing of solar cells.

To minimize the damage to the environment, the PV industry has introduced 'waste minimization programme' which includes recycling stainless steel cutting saws, recovering SiC from waste slurry and in-house neutralization of acid and alkali solutions.

Amorphous silicon: Some of the dangers faced in manufacturing crystalline silicon cells are present in the case of a-silicon also. The main hazard arises from silane gas (SiH4), when handled in large quantities. It is extremely flammable and explosive which ignites spontaneously in concentrations above 4.5%. In small quantities, handling is reasonably safe. With the a-Si production being projected to more than 100 MW scale, the risks will be multi-fold.

CdTe solar cells: Pure cadmium is a highly poisonous element produced as a byproduct of zinc production. The problem lies with the disposal methods. Cadmium is toxic even at low concentrations leading to serious health problems. The usage of cadmium is highest in the manufacture of paints, pesticides, stabilizers and batteries, and its use in PV industry is limited at present. It is essential to monitor cadmium levels in a CdTe cell plant to ensure that the levels do not reach chronic poisoning stage. Cadmium in CdTe form is stable because it is bound to tellurium and hence safe. The amount of cadmium in CdTe module is far less compared to that in a Ni-Cd battery often used in homes. A proper disposal of CdTe solar cells/modules ensures prevention of cadmium contamination of the environment. Recycling of modules to recover cadmium is the best approach and it is practiced in PV industry.

The cadmium used in PV is that extracted from the waste generated from mining and production of zinc. If Cd is not extracted and used in PV, the waste would be disposed off in the same quantities in a way that affects environment. Experiments have shown that even in the case of accidental fires, CdTe would be captured in the molten glass and would not be released into the environment (Ref. 38d). In any case, the cadmium that could be released into environment by CdTe cells in a worst case are estimated at around 0.02 gm/Gwh (Ref. 38a b c). Raugei (Ref. 38 d) has assessed the status of CdTe technology, 50 years into future, in relation to the environmental impact of Cd use and emission flows. The analysis is based on three alternative scenarios for Cd flows in the world and in the EU-27. The results clearly establish that, while CdTe technology may account for nearly one third of the global Cd demand in 2050 in each of the scenarios, Cd emissions would still be negligible, when compared to other existing sources. This result rules out any possible Cd-related environmental pollution coming from CdTe technology.

CI(G)S solar cells: The use of hydrogen selenide (H_2Se), a poisonous chemical, in the production of CIS cells causes concern. Even at a concentration of 1ppm, it is considered as 'Immediately Dangerous to Life and Health'. Its presence may lead to chronic poisoning too.

To protect people from exposure, the deposition system is enclosed under negative pressure and all exhaust gases are passed through an emergency control scrubber. The people are monitored to ensure they are not being exposed to the level of poisoning.

GaAs cells: Originally, these cells are designed for space applications because of their reliability. Now, they are being used in concentrator systems for terrestrial applications.

In the production of these cells, metal organics, hydrogen, and toxic hydrides gases such as arsine (AsH_3) and phophine (PH_3) are used. All of them pollute the environment. But the risks are minimal because smaller quantities of materials are enough to prepare cells for use in concentrator PV systems compared to the usual flat plate modules. However, the environmental performance could be improved by improving materials efficiency in production and reuse of GaAs wafer. Improving the energy efficiency of MOCVD process also results in higher environmental performance.

The common concern with all PV solar cells is 'leaching' if the cells/modules are not properly disposed. The standard test set up by Environmental Protection Agency (EPA) of USA for leaching is not followed by most of the manufacturers at present.

The 'Waste Electrical and Electronic Equipment' legislation, and the 'Regulation of Hazardous Substances' passed by European Union, which determines the substances that can be used in manufacturing and how they should be disposed, does not cover the PV industry at present. However, the industry is expected to take a life time approach to their products and provide for the proper collection and recycling of potentially hazardous components.

A few manufacturers have initiated steps for recycling and proper disposal. Deutsche Solar runs a programme for recycling of its poly silicon cells and modules. A voluntary take back system capable of growing to meet future waste recycling demands has been proposed by the company. For more details, refer to (Refs. 39, 40). Details are given by Wambach of Deutsch Solar (ref.39a). Several working groups have developed recycling concepts for thin film modules (Refs39 b,c,d). The recycling process which can be used for PV modules of any technology is discussed in Ref.39a.

The process of solar cell recycling is also discussed in detail by the author. The end of life costs of PV modules will significantly contribute to the overall costs of PV systems in the future with about 0.1 – 0.4 €/Wp. Soltech have developed a recycling process for silicon cells (Ref. 41). The reclaimed silicon wafers maintained their quality, providing high efficiency recycled Si solar cells. Their recycled wafers consumed 2.17 kwh/ wafer compared to 9.32 for virgin material. Sharp uses a system in which the silicon is remelted for recasting into ingots. First Solar conducts a take back and recycling programme of its CdTe modules. They have developed two recycling technologies for CdTe at the end of life modules and as manufacturing scrap prior to lamination. A process for recycling c-Si has also been developed (Ref. 42).

Solar World AG has a take back programme for its modules. To enable such recycling programmes easier, the companies are investing in mechanical innovations to make the dismantling process simpler (Ref. 43, 44). Several others also have come up with recycling processes. Drinkard Metalox of USA developed a process for recycling 'end of life modules' and manufacturing scrap of CdTe and CIS thin film modules (Ref. 45).

UNISON in collaboration with Brookhaven National Laboratory has reported recycling of CdTe modules (Ref. 46). Inter Phases of USA has developed a electrochemical approach to recycle CIS modules (Ref. 47), in which thin films are dissolved from the substrates and then the semiconductors are separated and retained.

Malinowska et al (Ref. 48) have recycled and reused chemicals from baths used for the chemical bath deposition of CdS. This process is simple and easily controllable to upscale for industrial use. Dhere et al (Ref. 49) have also studied CdS bath recycling and toxic waste reduction utilized in the production of CIS cells.

Take back schemes, designing products for easy recycling, developing efficient methods of recycling, and phasing out of hazardous substances wherever possible should be successfully investigated to make PV industry as an environmentally responsible one.

Reuse of Silicon:

The PV solar industry is rapidly growing and the demand for silicon feedstock is on the increase. Several approaches to produce solar grade silicon on large scale have been discussed in Chapter 2. However, better use of off-grade silicon, and reclaim of suitable by-products along the whole PV value chain also contribute to the silicon feedstock. Even from the environmental and economical considerations, the reuse of solar silicon by-products like cut-offs of crystals, ingots or bricks, broken wafers and process failures are essential. The availability of off-grade silicon from the semiconductor and PV industry and the possibility of its reuse in high yields, and the related mechanical and chemical processes are discussed by Wambach (Ref.40). His analysis reveals that the reuse of silicon contributes up to 40% of the silicon demands in an environmentally friendly and cost-effective way. The need for further research to devise better methods for quality inspection of recycling material has been suggested.

REFERENCES

1. ***R.W.Miles et al :*** *Prog.Crystal Growth and Charact. of Materials, 51, p 1 (2005)*
2. ***H.Fischer:*** *Proc. 1st ECPV Solar Energy Conf., Luxembourg,p52 (1977)*
3. ***M.Wolf et al:*** *Proc. 13th IEEE PV Specialists Conf., Washington, p271 (1978)*
4. ***K.Hay et al:*** *Proc. 15th IEEE PV Specialists Conf., Kissimme, p267 (1981)*
5. ***F.Riddoch et al :*** *Solar Cells, 2, p141 (1980)*
6. ***G.Hagedorn et al:*** *Proc. 9th EC PV Solar Energy Conf., Freiburg, p542 (1989)*
7. ***W. Paltz et al:*** *Intl. Journal of Solar Energy, 10(3-4), p211 (1991)*
8. ***G. Jenkins et al:*** *WESC E/5A/CON/1632/172/099, Final report, Sec.2 (1981)*
9. ***W.H.Bloss et al:*** *Proc. ISES Silver Jubilee, Atlanta (1979)*
10. ***K.M.Hynes et al:*** *Proc. 10th EC PV Solar Energy Conf., Lisbon, p461 (1991)*
11. ***K.M.Hynes et al:*** *Proc. 1st World Conf. on PV Energy Conversion, Hawaii, p958 (1994)*
12. ***E.A.Alsema:*** *Renewable and Sustainable Energy Reviews, 2, p387 (1998)*
13. ***E.A..Alsema:*** *Report no.96074, ISBN 90-73958-17-2 (1996)*
14. ***E.A.Alsema et al:*** *Proc. 2nd World Conf. and Exhibition In PV Solar Energy Conversion, Vienna p2125 (1998)*
15. ***E.A.Alsema et al :*** *Proc.19th European PV Solar Energy Conf., Paris, p840 (2004)*
16. ***E.A.Alsema et al:*** *Proc. 20th European PV Solar Energy Conf., Barcelona (2005)*

17a. ***K.E.Knapp et al:*** *Proc. 16th European PV Solar Energy Conf., Glasgow ,p2053 (2000)*

b. ***K.Knapp et al :*** Solar 2000 Conference, Madison,WI, (2000)

18a. ***G.Peharz et al:*** *Progress in Photovoltaics: Res. Appl.*

b. E.A.Alsema : Proc. Workshop on Photovoltaics and the Environment, Keystone,CO (1998) & Solar FAQs, NREL document, DOE/GO-102004-2040, Dec.2004

19. ***M.Raugei et al:*** *Proc.20th European PV Solar Energy Conf. Barcelona (2005)*

20. ***K.M.Hynes et al:*** *Proc.13th European PV SE Conf., Nice, p2202 (1995)*

21a. ***K.Kato et al:*** *Prog. Photovolt. Res. Appl.6, p105 (1998)*

b. ***J.Mason:*** Report prepared for Tucson Electric Power, (2004)

c. ***E.A.Alsema et al :*** Proc.21st European PVSEC, Dresden, (2006)

[www.epia.org/fileadmin/EPIA_docs/publications/epia/EPIA_SG_IV_final.pdf]

d. ***R.Dones et al :*** Life-cycle Assessment of PV systems: Results of Swiss studies on Energy chains, Utrecht University, Report no.97072(1977).

22a. ***R Ottinger et al (Eds):*** *Environmental Costs of Electricity, Oceana Publs. Inc. New york (1991)*

b. ***E.A.Alsema et al :*** Energy Policy, 28(14), p999 (2000)

23. ***B. Sorenson:*** *Life Cycle assessment of Energy Systems, Report prepared for OECD (1992)*

24. ***B. Sorensen:*** *Proc.2nd World Conf. & Exhibition on PV Solar Energy Conversion, Vienna, p3461 (1998)*

25a. ***A.E. Baumann et al:*** *Proc.10th EC PV Solar Energy Conf., Lisbon, p834 (1991)*

b. ***European Commission, External Costs*** – Research results On socio-environmental damages due to electricity and Transport. http://www.externe.info

c. ***M.J.de Wild-Scholten et al:*** Proc. Workshop on 'Life-cycle Analysis and recycling of solar Modules – The Waste challenge, Brussels (March 2004)

26a. ***E.A.Alsema et al:*** *Proc.11th European PV Solar Energy Conf. Montreaux, p995 (1992)*

b. ***K.Komoto et al :*** Proc.23rd European PVSEC, Valencia, p3833 (2008)

c. ***M.Ito et al :*** 17th International PVSEC, Fukuoka (2007)

d. ***M.Ito et al :*** Prog. Photovolt: Res.Appl. 16, p17 (2008)

e. ***N.Jungbluth et al:*** Proc.23rd European PVSEC, Valencia, p3688 (2008)

27a. ***B.C.W. van Engelenberg et al:*** *Proc.12th European PV Solar Energy Conf. Amsterdam, p305 (1994)*

b. ***E.A.Alsema:*** Sustainability of PV systems on the basis of Advanced silicon technology, Utrecht University (2003)

c. ***E.A.Alsema et al :*** Proc.19th European PVSEC, Paris, p7-11 (2004)

28. ***K.M.Hynes et al:*** *Proc.12th European PV Solar Energy Conf. Amsterdam, p309 (1994)*

29. ***G.J.M. Phylipsen et al:*** *Environmental life-cycle Assessment of multi-crystalline Si solar modules, Report no.95057, (1995)*

30. ***P.Frankl et al:*** *Proc.19th European PV Solar energy Conf., Paris, p3373 (2004)*

31. ***Progress in Photovoltaics:*** *Research. Appl. 6, p87 (1998)*

32. ***V.M.Fthenakis et al:*** *Proc. 31st IEEE PV Specialists Conf. Orlando, p1655 (2005)*

33. ***V.M.Fthenakis et al:*** *Prog. Photovolt. Res. Appl. 3, p295 (1995)*

34. ***K.M.Hynes et al:*** *Proc.19th European PV Solar Energy Conf. Paris, p2529 (2004)*

35. ***H. Steinberger*** Proc. 2nd World Conf. and Exhibition PV SE Conversion, Vienna, p2276 (1998)

36. ***V.M.Fthenakis:*** Proc.19th European PV Solar Energy Conf. Paris, p2166 (2004)

37. ***V.M.Fthenakis:*** In 'Practical Handbook of Photovoltaics Fundamentals and Applications, p854 (2003), T.Markvart & L.Castaner (Eds)

38a. ***V.M.Fthenakis:*** Renewable and Sustainable Energy Reviews, 8, p303 (2004)

b. ***M.Raugei:*** Proc. 23rd European PVSEC, Valencia, p2584 (2008)

39a. ***V.M.Fthenakis :*** 'PV News Round up from the 33rd IEEE PV Event', news item in 'Printed Electronics World', May (2008).

b. ***K. Wambach:*** Proc.19th European PV Solar Energy conf. Paris, p2018 (2004)

c. ***R. Geganwart:*** Workshop on Life-cycle Analysis and Recovering of solar modules – The Waste Challenge, Brussels (2004)

d. ***N.Warburg:*** ibid

e. ***L.B.Geise et al:*** Proc. 1st International Energy and Environmental Symposium, Turkey (2003)

40. ***K. Wambach;*** Proc.19th European PV Solar Energy Conf., Paris, p572 (2004)

41. ***L. Frisson et al:*** Proc. 16th European PV Solar Energy Conf. Glasgow, p2160(2000)

42. ***J. Bohland et al:*** Proc.2nd World Conf.and Exhibition PV SE Conversion, Vienna,p716(1998)

43. ***A Cameron:*** renewable Energy World, 8(6), p58 (2005)

44. ***Mariska de Wild-Scholten:*** Proc.20th European PV SE Conf. Barcelona (2005)

45. ***R.E.Goozner et al:*** Proc. 26th IEEE PV Specialists conf., Aaheim, p1161 (1997)

46. ***C.Eberspacher et al :*** Proc.1st World conf. on PV Solar Energy conversion, Hawaii, p962 (1994)

47. ***S. Menezes:*** Proc. 2nd World conf.& Exhibition on PV SE conversion, Vienna, p597 (1998)

48. ***B. Malinowska et al:*** Proc.19th European PV Solar energy Conf. Paris, p1820 (2004)

49. ***N.G. Dhere et al:*** Proc.19th European PV Solar Energy Conf. Paris, p1966 (2004)

CHAPTER TEN

MARKET FOR PHOTOVOLTAICS

10

MARKET FOR PHOTOVOLTAICS

Solar PV technology is one of the renewable energy technologies which has attracted the attention of the policy makers in the development of sustainable energy strategy in the recent decades. The PV technology has some unique advantages over the other technologies, though it is not appropriate in all situations due to certain disadvantages.

10.1 ADVANTAGES AND DISADVANTAGES OF PV

The numerous advantages of this technology are:

- require no fuel, only sunlight which is free, copious, continuous and inexhaustible
- flexible source of small amounts of electric power; and can be produced from milliwatts to megawatts at the place of demand while avoiding transmission
- environmentally benign, emitting no pollutants and free noise, sustainable power as a means of reducing global warming,

- modular in nature, can be expanded to meet growing needs or dismantled and used for different applications or at different sites,
- cost-effective power for people living far off from the utility grid,
- reliable, durable, robust and weather proof prefered power option for critical applications that require consistent, predictable energy supply such as for health care, millitary communications; have life times of 15 to 25 years,
- no moving parts, so less likely to break down,
- light weight and can be easily transported and installed,
- requires low maintenance, making them popular power sources in isolated places-occasional cleaning of the module surface, checking the electrical wiring and topping up the batteries,
- produce DC electricity and can charge storage batteries,
- comprises many components which can be manufactured locally.

The drawbacks of the technology are:

- unsuitable for all cases of power requirement,
- high initial cost
- requires additional equipment to convert DC into AC
- infrastructure for maintenance and spares are inadequate in many countries

10.2 APPLICATIONS OF PV

Solar photovoltaic technology was a marginal technology before the nineties with limited applications. The major applications of solar PVs before early nineties have been for

* *consumer products such as watches, calculators, cameras, toys and so on;*
* *satellites and*
* *remote area power supplies.*

From early nineties, solar home systems (SHS), lighting systems and building integrated PV systems have mainly contributed to the development of PV market globally. Extensive research and development activities especially in USA, Europe and Japan as well as increasing manufacturing facilities have brought down the costs significantly and contributed to the growth of the PV market.

(1) Satellites:

The first major application of PV technology was probably in satellites because of its reliability in supplying power continuously. Satellites require very reliable lightweight and uninterrupted power, and PV electric power is the ideal source because the technology depends on the Sun for enrgy. Though the PV power systems were highly expensive during fifties, their reliability to power the satellites justified their use.

With four decades of intensive research, the efficiency to weight ratio of the PV panels outweighs the economic considerations of today. The area of satellites has become a significant market with many countries entering into space programmes.

(2) Telecommunications:

It is a service sector essential to improve the quality of life in remote and rural areas. Due to the high cost involved in providing grid power to the remote areas, the telecommunication facility has not penetrated much into those areas. Rapid and accurate information exchange is important for the success of health services, education, transportation, agriculture, energy management, water management, disaster relief, government, industry and trade, social services and economic development. Photovoltaic technology has provided a cost effective solution to this problem. PV power has been used in several countries to power relay and repeater stations and also telephones. Several thousands of these systems have been installed in many developing countries which have proved to be highly reliable and relatively low cost for operation and maintenance. In India, PV power systems for radio telephones have been popular with more than 190 thousand systems supplied up to the year 2000. It has been one of the largest markets for PVs with continuous growth. Similar principles apply to solar powered radios, television sets, emergency telephones and remote monitoring systems(RMS). The RMS are used to collect weather data and information concerning environment.

(3) Cathodic Protection:

It is a method used to protect metal structures from corrosion. This requires a small negative DC voltage applied to the metal structure which prevents it from oxidizing and rusting. The positive end of the source is connected to a sacrificial anode which is generally a piece of metal which corrodes. This small negative voltage is supplied by PV. The required protection stretches generally long distances as in the case of pipelines, bridges and railway lines. The other structures are buildings, tanks and wells. This has been one of the fastest growing applications of photovoltaics.

(4) Water Pumping:

This is another ideal application of photovoltaic technology. There are more than a few tens of thousands of water pumps powered by solar energy around the globe, especially in developing countries. These are used for irrigation, for supplying water to livestock, for village water supplies and for domestic consumption. Most of these pumps are used to lift water from wells, streams and rivers. A typical solar powered pumping system comprises a PV array and a dc pump, with no energy storage facility. Depending on the application, either a DC submersible pump or DC surface pump is used. For the former, a controller is required which interfaces the PV array and the pump. The controller enables to maximize the daily water delivery while protecting the power source and the pump. Two types of controllers are available,- current booster type and voltage booster type and the selection of the controller depends on the situation.

Solar powered water pumps with large capacity are widely used in some African countries for domestic consumption. In India, this programme has been under operation for nearly ten years with subsidy incentives from the federal government. PV array of about 1 kWp capacity along with a DC pump are used in the Indian programme, mostly to lift water from rivers and lakes to irrigate small pieces of land. They are also suited for drip irragation or micro sprinklers.

In Gambia, a 4 kW PV powered pumping system uses sun trackers and electronic controls to increase system efficiency, reducing the number of required PV modules. This system pumps up to 30000 gallons per day for irrigation. In northwestern Kenya, in extremely remote desert areas, 2 kW PV pumping system mounted on trackers provides about one thousand gallons per day to a population of about 100 tribals and about 250 cattle head.

In Haiti and Mexico, very simple PV systems using fixed PV arrays are used to power small pumps.

The largest manufacturer of pumps in the world, Grundfos offers solar-powered submersible pump (model:SQFlex), which is extremely durable and requires very little maintenance. This model can be operated with solar power, wind power or a combination of both. The latter option is possible due to the special feature of the pump, namely, the input can handle both alternating and direct currents without any manual switching because an inverter with MPP tracking is integrated into the pump motor. The heart of the unit is a permanent magnet motor, which operates independently of the frequency. The pump is designed for 30 to 300V DC and 90 to 240V AC; and the typical operating voltages lie at around 120V, for which three or four 55Wp PV modules with MPP voltage of approximately 34V each are needed. These are installed in many African countries, Australia, Thailand, and USA.

'Total Energie' of France offers a DC system (model:TSP1000) comprising six to twelve 50Wp modules, a pump with brushless 4-inch submersible motor and a controller. The maximum head is 70 meters and the maximum delivery rate is 13000 litres per day. These pumps are operational in Brazil, Mali, Ghana and Burkina Faso.

Grundfos and Total Energie supply AC pumps too, which are preferred when high pump heads and delivery rates are involved. AC pumps are less expensive and require low maintenance. The inverters for AC pumps usually supply 3-phase current.

Bernt Lorentz AG of Germany brings diversity into the DC submersible pumps. The most powerful version raises water from a depth of 240 meters delivering up to 4000 litres/hour; and the mini model delivers up to 2000 litres/hour with a maximum head of 50 meters.

Wurth Solergy of Germany offers both DC pump and AC pump. The DC pump, depending on the configuration, can deliver up to 20000litres/day with heads up to 160 meters. The more powerful AC pump uses up to 16 PV modules of 110Wp each with a special inverter and is optimized for mid-range heads and delivery rates. For example, at a depth of 25m, it delivers up to 42000 litres/day. These are installed in many African countries for PV based water supply in villages.

(5) Water Treatment:

In remote areas and in villages too, photovoltaic power is used to purify drinking water. PVs are used to power a ultra violet light used to kill bacteria in the drinking water. Generally, the water treatment system is combined with a solar powered pumping system. Desalination of brackish water is also done using PV power. Many countries utilize solar powered desalination plants to obtain potable water.

(6) Electric Fencing:

Electric fences are widely used to protect agricultural crops from wild animals specially in remote regions bordering forest ranges. These fences have one or two 'live' wires that are maintained at about 500 volts dc, which gives a harmless but painful shock to any animal that touches them. These fences are used in security areas and wild life enclosures. The requirement is a high voltage at very small current; hence it can be met by a photovoltaic system consisting of PV array, storage battery and power conditioning equipment. This is another potential market for PVs in developing countries.

(7) Lighting Systems:

The cost of power required for lighting especially in remote areas is too high to use the grid because the requirement of power is not high. In these areas, kerosene lamps, candles or conventional battery-powered lights are used to obtain light. PV can be a more economical option compared to these services, and at the same time provide clean, reliable and plentiful light. In addition to house-hold lighting, the remote lighting includes security lighting, road signs, railway crossing signs, navigation aids, village street lighting . The PV system for these purposes consists of a PV module of 35 to 50 Wp capacity, a small storage battery, a power conditioner and a low-voltage high efficient dc fluorescent lamp. In developing countries, these lighting systems are very convenient; in India, in the last 15 years, a few hundred thousand solar street lighting systems have been installed in many rural dwellings. In Brazil, Mexico, Dominican Republic, Honduras and some African countries, these PV lighting systems are installed in rural areas as a less costly alternative to the extending of grid.

Solar lantern is solar powered portable lighting system which is considered to be a viable and superior substitute for a kerosene lantern used in rural households in many countries. The Solar lantern consists of a 5 or 7 watt compact fluorescent lamp (CFL), a 12V 7AH battery and a small controller. The lamp is powered by a 8 to 10 Wp PV module and provides light for 3-4 hours per day. The PV lantern has found wide acceptance because the problems associated with kerosene lantern such as smoke, emission of CO_2, possible fire hazard, uneven and poor quality of light are eliminated.

The light output from PV lantern is not only even but very much higher, around 380 lumen compared to about 68 lumen from the kerosene lantern. Though the initial cost of a PV lantern is high, considering the kerosene costs (about 100 lt, per year on average), the NP (net present) value of the life-cycle cost for the output delivered, PV lantern appears to be highly preferable over the kerosene lamp.

(8) Rural Electrification:

Photovoltaics have the potential to meet the power requirements of about two billion people in the rural settlements long before they can be served by the grid (Ref. 1). PV power systems have proved to be a low cost alternative to extending grid lines. These systems are used to provide low voltage electric power for lighting, communications, vaccine refrigeration, sanitary services, educational purposes and battery charging which are the basic requirements of the rural people. Vaccine refrigeration is an important application; vaccines are very sensitive to temperature fluctuations and must be maintained between 32 and 46 F (0 to 8C) to retain their potency. This becomes a formidable task in areas with no grid power. International health organizations have incorporated PV into their Vaccine delivery system due to their reliability and temperature stability. Thousands of PV systems currently operate in rural health facilities located in remote areas in several countries of Africa, South America and Asia. These systems maintain critically needed serums and supply power to medical equipment.

In many Asian and African countries such as India, Bangladesh, Indonesia, , Kenya, Gabon, Vietnam, South Africa, and so on, these village PV systems are installed extensively under various national and international programmes to improve the living conditions, to stimulate local economies and to halt migration of people to urban areas. For example, in Gabon (East Africa), a programme to install PV systems to 100 villages has started last year to provide power for medical stations, schools and houses in the country's remote villages. The school systems have a peak power rating of 660 w, the medical stations, 550w and for domestic and street lighting applications, 110 w. Siemens, Germany, supplies these systems as turn-key installations to provide power for lighting, fans, vaccine refrigeration and communications (Ref.2). In Mongolia, rural electrification of 100,000 nomadic homes has started to improve basic standards of living and improving social welfare of rural population. This programme is expected to be completed by 2010 (Ref. 3a). This offers a potential market for PV.

(9) Solar Home Systems:

In the solar home system (SHS), the electric power is produced on the roof of the house for the individual family. For a typical rural house, the load consists of a couple of lights and a television or radio; hence, a PV system of 50 to 75 Wp which can provide a daily average yield of 180 to 270 Wh is adequate. These are also extensively, (over one million), installed in many villages of Asian, African and Latin American countries . In India alone, more than 117,000 home lighting systems are installed by 2000 which is a fast growing market for PVs. Of all the applications, SHS have found to be the most successfully accepted in both urban and rual India because the grid power is not available when needed most of the time. 5000 solar home systems, each of 10 to 30 watt capacity, have been installed in Mongolia just to provide lighting in nomadic homes. Around 1,50,000 in Kenya, more than 100,000 in China, 85000 in Zimbabwe, 10000 in Indonesia and 40000 in Mexico have been installed.

Many models for SHS market are offered by a large number of manufacturers worldwide. The kits released by a few selective manufacturers in recent years are briefly outlined (Ref.3b). These systems are operating in many countries of Africa, South America and Asia.

'Free Energy Europe' of Netherlands, released three different configurations of the System to provide lighting, to operate a TV and to charge cellular phones. These Systems use 14 Wp amorphous silicon models of their own make. The special feature is a plastic frame which sits directly on the glass and produces a water-tight seal. Fluorescent lamps from Indonesia and charge regulators from Germany are obtained. A double switch for day-time and night-time lighting, and a cable for charging cellular phone are also provided. These kits are marketed on a large scale in East African countries, especially in Kenya.

'Steca' of Germany offers a portable SHS with a system voltage of 12V for DC consumers. A lead acid battery of 7Ah capacity linked to a charge controller is provided. An integrated DC/DC transformer with outputs of 3,6,9 volts provide connection to a small radio or cassette player. The kit operates over a temperature range of 0 to 40C. The manufacturer claims that there is no electronic discharging of the battery and the stored energy is maintained over several months.

'Total Energie' of France came up with a SHS comprising 50 Wp module, four lamps and a 12 volt output for operating television/radio/phone charger. Kits with 75Wp and 100Wp are also available. A large number of these units are installed in Morocco.

'Sharp Electronic Europe' of Germany, a newcomer to SHS market produces small systems with an AC output also. The kit consists of a 80 or 123 Wp PV module, a battery, connecting cables, a charge regulator and an inverter. The installation is simplified by connecting the module and battery to the compact unit. The daily power yield is at 200 Wh for the 80 W module, and 300Wh for the 123 Wp module as specified by the manufacturer.

'Solar-Fabrik' of Germany produced a System which provides an AC output of 200 watts at 230V, as well as 12V DC output. The main component of the System is the vario unit comprising a charge regulator/inverter combination, a solar battery of 90 Ah capacity and DC and AC outputs. The system comprises one or two 115 Wp modules depending on the load. The system specifications say that one module is sufficient to operate a colour television or computer.

SHS could fit in excellently for the Kyoto Protocol's Clean Development Mechanism (CDM) because, they directly displace green house gas emissions while contributing to sustainable rural development. SHS is a highly growing market in developing countries.

(10) Remote Houses:

Remote houses in the developed countries create a sizable market for PV. It was estimated in 1987, that 70000 remote houses in European union would be electrified using PV technology with less cost than with grid extension (Ref. 4) because of the exorbitant cost involved in running transmission lines. As the PV technology costs have been on the decrease in the recent years, the tendency to use PV in the remote houses in Europe has been on the increase. To cite an example, in Finland, 1-module PV systems to power lights and TV, 2-module systems to power TV, lights and refrigerator, and multi-module systems for higher power needs are installed (Ref.5).

(11) Grid-connected Systems:

The small-scale grid connected systems represent a radical change from the traditional concept of centralized electricity distribution. The generation is distributed and sited close to the local loads, enabling enormous savings in utility operations. By 2010, the grid connected market, both small-scale and large-scale, is expected to become economically viable. Building integrated PV has been one of the biggest applications of the grid connected systems, which has been detailed already.

Another application of small-scale distributed generation is to boost the power along an electricity transmission lines. When the distribution line is too long, the loads along the line can result in poor power quality. By installing PV systems along the line, a utility is able to support voltage and reduce transmission losses. The upgrading equipment required to boost the voltage is highly capital intensive. In certain conditions, providing PV power to a substation at a critical point to boost the power quality may be cost-effective than providing upgrading equipment. But as the costs come down, this application will become a significant market in many countries.

The large scale grid connecting systems are the central power plants built on 'green' fields, apart from BIPV. Such systems are considered as the ultimate application of photovoltaics. Some utilities are installing central power stations and generating MWp-scale power under the support of federal governments. Many such plants are operating mostly in European countries and USA, providing not only reliable power to utility grids but installation and operation experience and a realistic picture of the system economics and performance.

Large-scale grid-connected plants have come up since early eighties. The plant at Hesperia of 1 MWp capacity, at Carissa plains of 5.2 MWp capacity and at Rancho Seko of 1 MWp capacity are the ones installed in USA during eighties. In Europe, large-scale grid connected systems have appeared during nineties: 1 MWp in Toledo, Spain in1994, 3.3 MWp in Serre< Italy in '94, 1 MWp in Vast, Italy in '93, 5 MWp in Crete, Greece in '97. The Crete plant is a part of a proposed 50 MWp system funded by European Union and Greek government, using poly-silicon modules.

These installations provide a hope that central PV power generating plants can be developed and promoted with economic viability in future.

(12) Miscellaneous Applications:

PV is utilized in many other applications: consumer products, emergency power systems, vaccine and blood storage refrigerators for remote places, portable power supplies at camping sites, navigational aids, road signaling, environmental monitoring, parking meters, highway telephones, car sunroofs, ticket machines, aeration systems for ponds, fountains for garden ponds, news paper dispensers and so on.

10.3 GLOBAL PRODUCTION AND COSTS OF SOLAR PV

10.3.1 Introduction

The solar PV market is booming. In recent years, Solar PV has been emerging as a mature player in the energy sector. The annual growth rates in the last five years are in average greater than 40%. The global PV shipments from 1990 are shown in Figs 10.1 and 10.2 (Ref: H. Ullal, NREL).

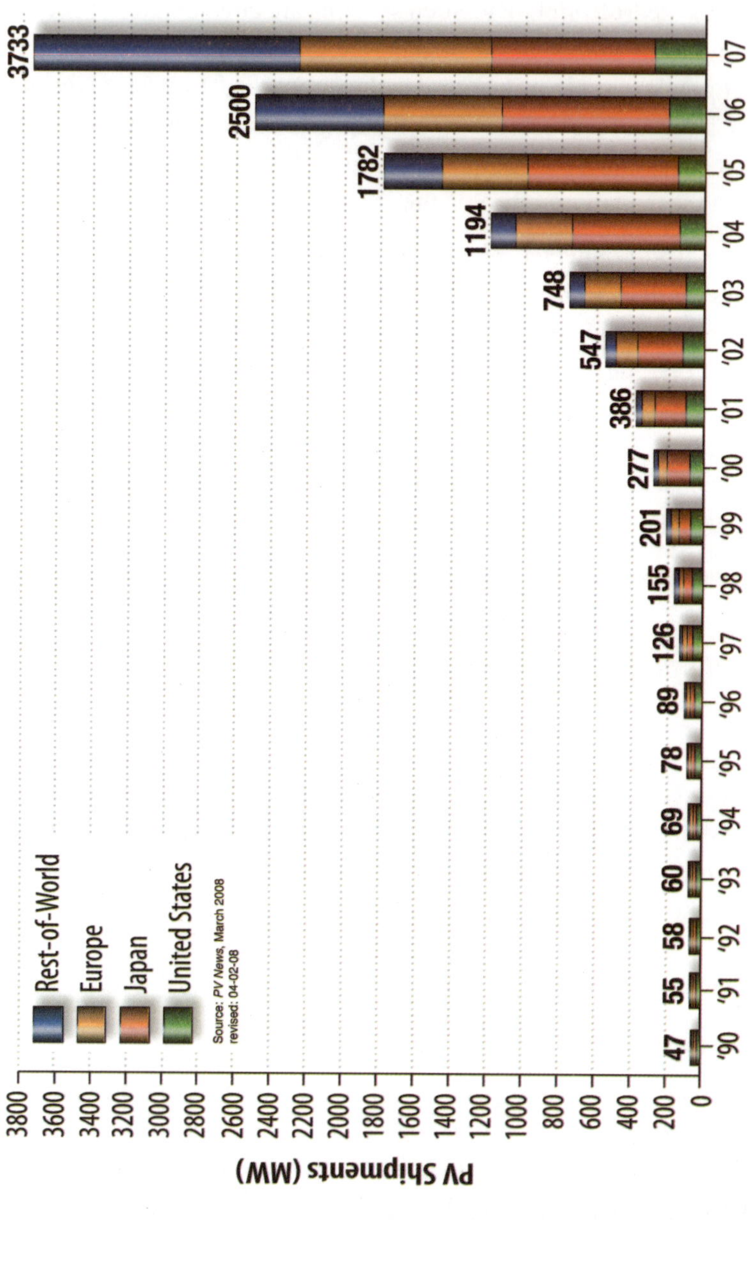

Fig. 10.1 *Global PV shipments*

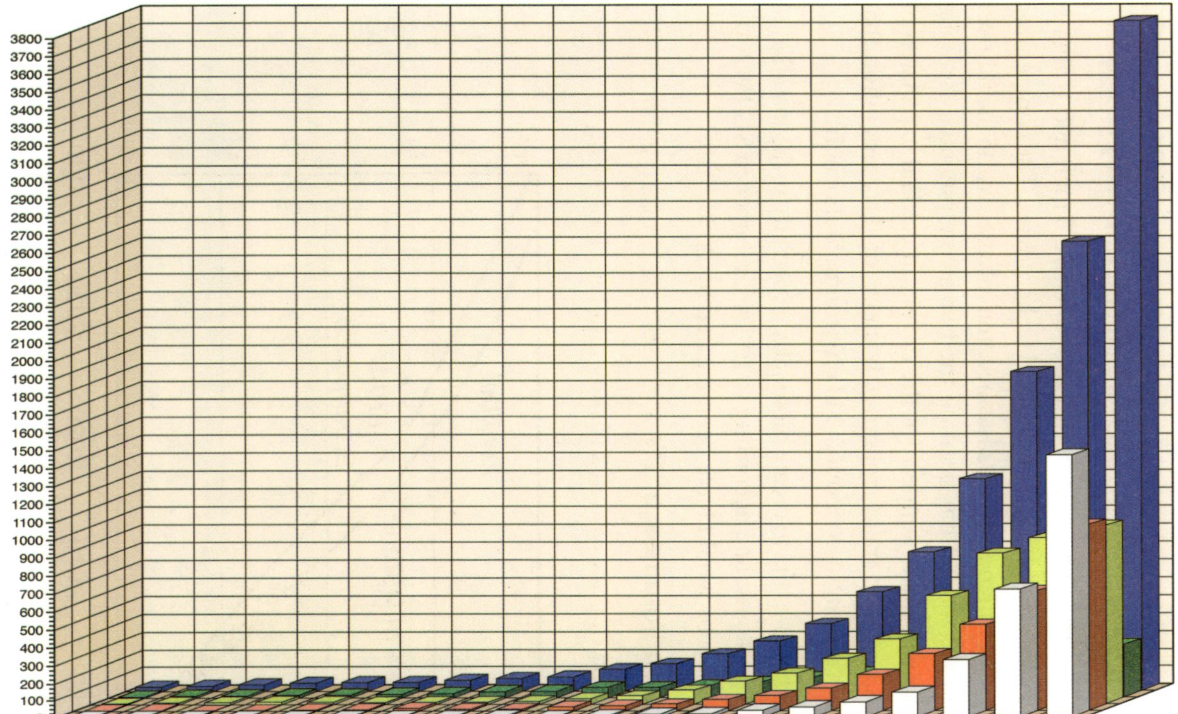

	1988	1989	1990	1991	1992	1993	1994	1995	1996	1997	1998	1999	2000	2001	2002	2003	2004	2005	2006	2007
Rest of World	3	4	4.7	5	4.6	4.4	5.6	6.35	9.75	9.4	18.7	20.5	23.4	40.6	53.3	81.3	141.5	322.5	714.0	1484.1
Europe	6.7	7.9	10.2	13.4	16.4	·16.55	21.7	20.1	18.8	30.4	33.5	40	49.8	73.9	122.1	200.2	311.8	472.8	657.3	1062.8
Japan	12.8	14.2	16.8	19.9	18.8	16.7	16.5	16.4	21.2	35.0	49	80	128.6	171.2	251.1	363.9	601.5	833.0	927.5	920.0
United States	11.1	14.1	14.8	17.1	18.1	22.44	25.64	34.75	38.85	51.0	53.7	60.8	75	100.3	120.6	103.0	138.7	154.0	201.6	266.1
Total	33.6	40.2	46.5	55.4	57.9	60.09	69.44	77.6	88.6	125.8	154.9	201.3	286.8	386.0	547.0	748.4	1193.5	1782.3	2500.3	3733.0

From PV News, February/March editions.

***Fig. 10.2** Global PV shipments*

Since 20006, the growth has been much more impressive and is almost 50% annually, compared to earlier years. The shipments in 2008 were 5492 MW according to a communication from Northumbria University (2009).

The global PV industry, particularly in Europe, USA, Japan and China is investing heavily in new production facilities and technologies. The competition among established major manufacturers has become intense, while the number of new entrants into the market has been on the rise as the potential for solar PV opens up. The key political support for the development and expansion of solar PV is now extended in many countries because of the recognition that solar power would bring socio-economic and environmental benefits.

PV industry has the potential to create more jobs, especially in the installation and servicing of PV systems, and according to EPIA-Greenpeace SG IV 2007, more than 6.3 million jobs would have been created by 2030.

10.3.2 Production Volumes of PV

The annual production of photovoltaics is shown in Fig. 10.3. (Ref. 6)

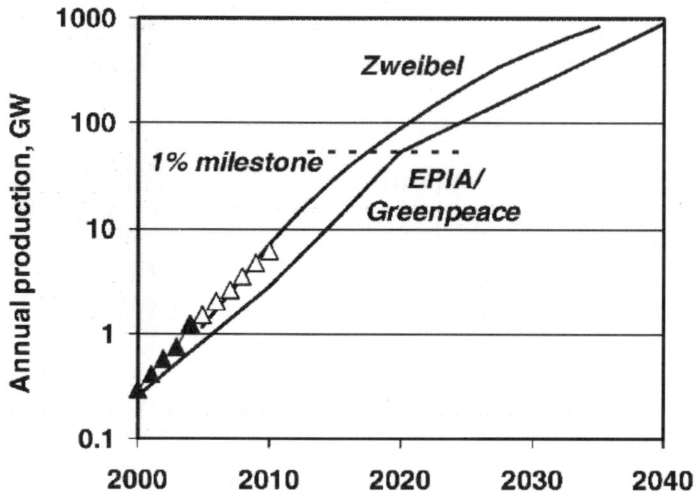

Fig. 10.3 *Annual production of photovoltaic modules showing actual data (▲) plus near-term (Δ) and long-term projections*

The actual production for 2000-2004 and the projections up to 2040 are given (Ref. 6). The lower curve is based on EPIA-Greenpeace study, which assumes annual growths of 27% during 2000-2010, 34% during 2010-2020, and 15% during 2020-2040. The projections in this study appear to be 'conservative' in the first few years because the actual productions turn out to be higher than the predicted.

Based on the recent rapid progress in PV, Zweibel of NREL (Ref.7) has estimated the production volumes for the next 60 years. He assumes annual growth rate of 38% around 2010, relaxing to 20% by 2020, to 10% by 2035, and settling at 5% by 2065.

Several other market predictions for '2010 PV market 'vary very much: 5.65 GW (Ref. 8), 7 GW (Ref. 9), 8.25 GW (Ref.10), and 10 GW (Ref.11). The SG IV 2007 further mentions that if the massive capacity increases underway or announced are all realized, the global production capacity of solar cells may exceed 20 GW in 2010/11. The projected growth rates, however, are subject to essential requirements: availability of sufficient silicon feedstock, a fast growing market for the PV product, and significant reduction in solar cell cost.

The EPIA - GREENPEACE Solar Generation IV-2007 Report (Ref. 8) and Swiss Sarasin Bank Solar Energy Report, 2007 (Ref. 10) are comprehensive reports on the current PV industry reflecting all the relevant aspects, alongwith a few considered projections.

10.3.3 Past & Present Costs of Wafered Silicon Solar Cells

Wafer-based crystalline silicon cell production still dominates and accounts for 91% of the PV module market. The main reason for this market leadership is its 'technology development' in the last three decades. The module price of $32 per watt in 1979 has come down by ten fold, to about $3 per watt due to increased volumes of production. The prices have followed a classic experience curve in Cost Vs Cumulative volume, and the behaviour is shown in Fig. 10.4 (Ref.12)

The experience factor is 80%, meaning that module prices reduce 20% for every doubling of the cumulative volume. The blunt during 2004 and 2005 in the historic trend is due to the shortage of silicon feedstock. Sufficient poly silicon is expected to be available from 2009 and if this happens, the historic price reduction is expected to continue.

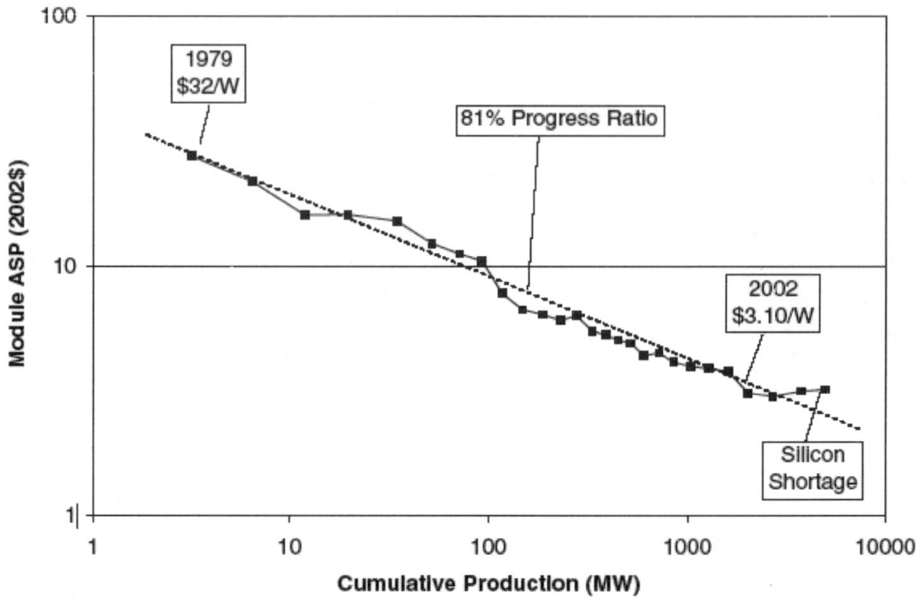

Fig. 10.4 *Historical plot of module price showing classic experience curve behavior. Data from Strategies Unlimited and Navigant Consulting. All prices are corrected to 2002 US dollars (Ref. 12)*

Stated below, are the factors which together contributed to the historic price reduction: (1) a decrease in the price of silicon feedstock from about $300/kg in 1979 to $30/kg in 2002, and an increase again to $40-50/kg for long term contracts. Many industry observers expect the long-term price to be $20/kg because many companies in several countries are investigating alternate routes to the standard Siemens process for the production of solar grade poly silicon feedstock. The silicon consumption per watt may also drop from the current 10 gms/watt to 7 to 8 gms/watt, thereby contributing to the decrease of the cell/ module cost, (2) wafer thickness which used to be around 500 μm in 1979 has gradually decreased to 300 μm by 2002. Several manufacturers are now making wafers of thickness 200 μm or less. Efforts are underway to bring down the thickness to 150 μm. This will significantly contribute to the cost reduction, (3) the cell size of 75 mm in 1979 has increased to the present 156 mm (e.g. Q-cells and Trinasolar) and is likely to increase further. With increase in cell size, the amount of handling for a given power decreases, (4) initial method of forming the metal grids by vacuum evaporation has been replaced by screen-printing which has greatly contributed to the cost reduction, though it has affected efficiency to an extent, (5) sawing of ingots to obtain wafers has been a standard practice from the beginning and with improvements over the years, the wire-saws currently can slice not only a large number (4000 wafers) at one time increasing the productivity, but slice thinner wafers reducing the kerf loss,(6) the process development and capacity expansion also contribute significantly to cost reduction and economy-of-scale cost benefit. Altering/introducing many processes in the production line, modifying the device structure, improving in-line process monitoring and output control in cell production have taken place over the years. The factory production capacity has risen from about 1 MW/year in 1980 to the present 100 MW/year which has become a general norm now. There are a few with higher capacities also, (7) automation in the production improves the yield and reduces the labour costs. Many present day manufacturers prefer fully automated production lines.

10.3.4 Projected Costs of Wafered Silicon Cells

A survey conducted among 23 silicon experts at an NREL workshop in 2002 has revealed that among several factors that are likely to contribute to the cost reduction in the coming years are the 'efficiency' and 'wafer thickness' which will have the highest impact (Ref.13). Figs 10.5 and 10.6 demonstrate the results. The module cost is computed from these results and shown in Fig. 10.7.

The projected module cost is \$ 1.00/watt in 2012 (Fig.10.7). The graph further indicates that the mono crystalline silicon costs will be lower than the multi-crystalline silicon because of thinner wafers and higher performance. It will be indeed a great achievement!

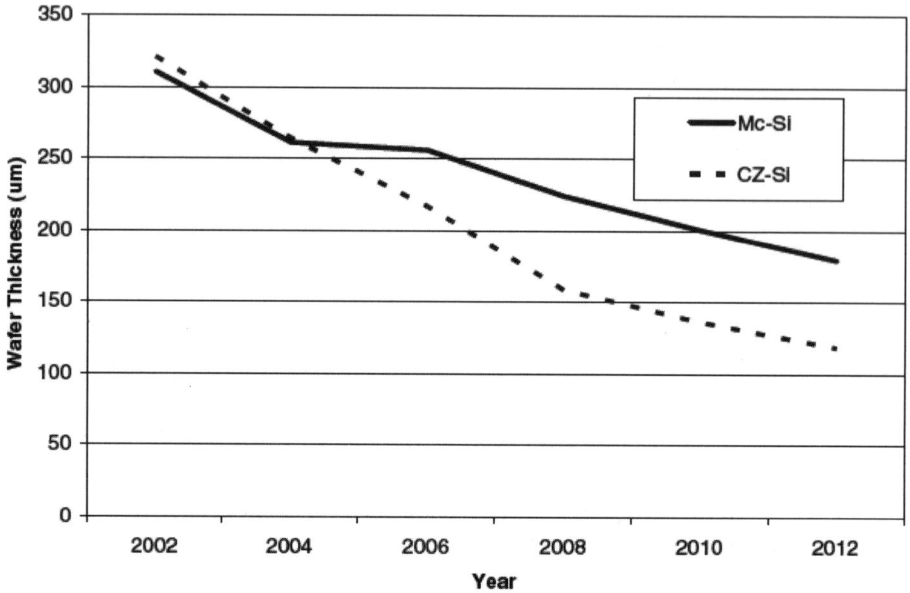

Fig. 10.5 *Projected wafer thickness (Ref. 13)*

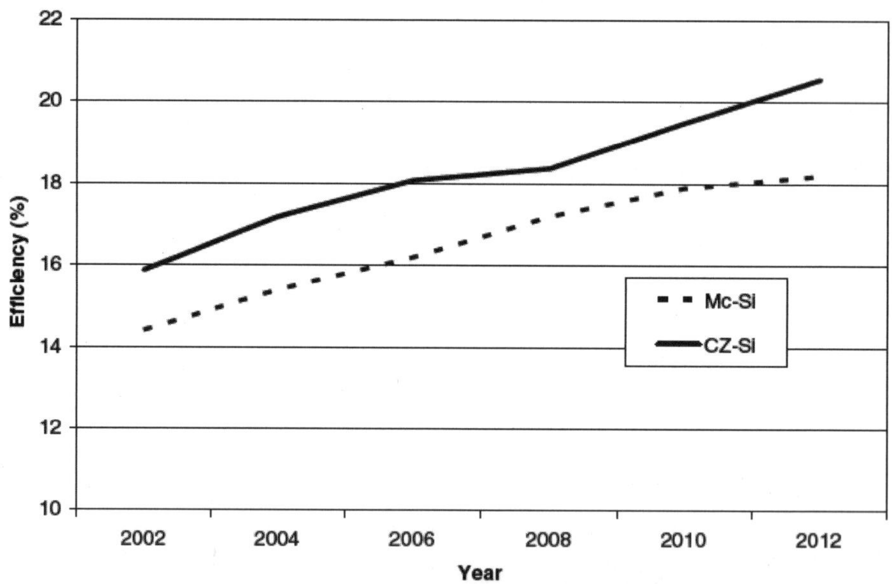

Fig. 10.6 *Projected cell efficiency (Ref. 13)*

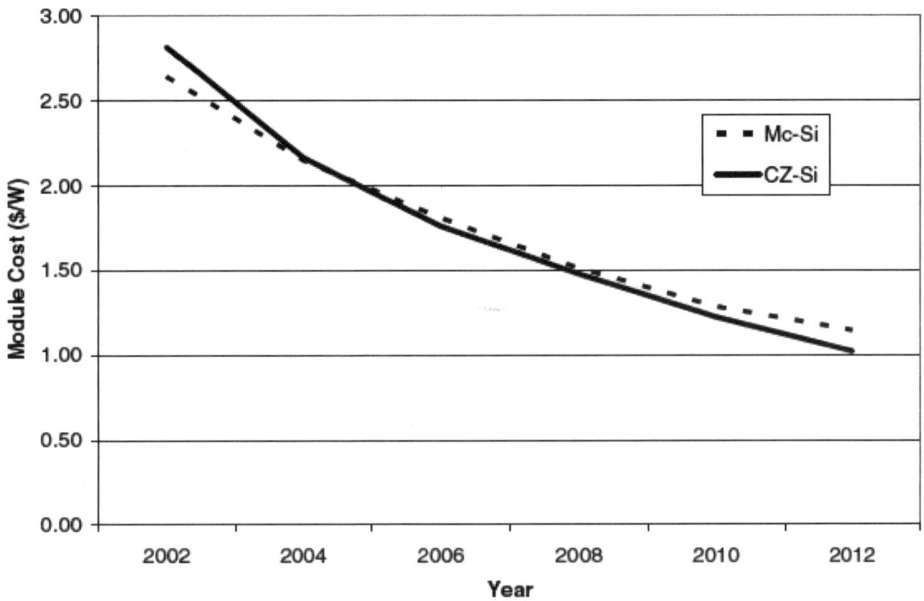

Fig. 10.7 *Projected module manufacturing cost (Ref. 13)*

The cost projections, however, call for a cell efficiency of 20% and wafer thickness of 120 μm by 2012. High efficiency concepts reviewed by Zhao et al (Ref.14) in 2005 will be relevant in this context. The other factors which the industry has to concentrate on are poly silicon production, crystal growth, sawing and kerf loss, and automation in production including steps for in-line quality assurance to achieve the projected cost.

Fig.10.8 shows the projected cost by extending the historical experience curve. Here, a continuous market growth of 30% per year is assumed. The graph projects a cost of $1.56/watt in 2012 as against $1.00/watt obtained in Fig. 10.7. Even with this price of $1.56/W, the system cost may come to around $3.00/watt, which may still be cost-effective for grid-connected PV in many Asian and African countries.

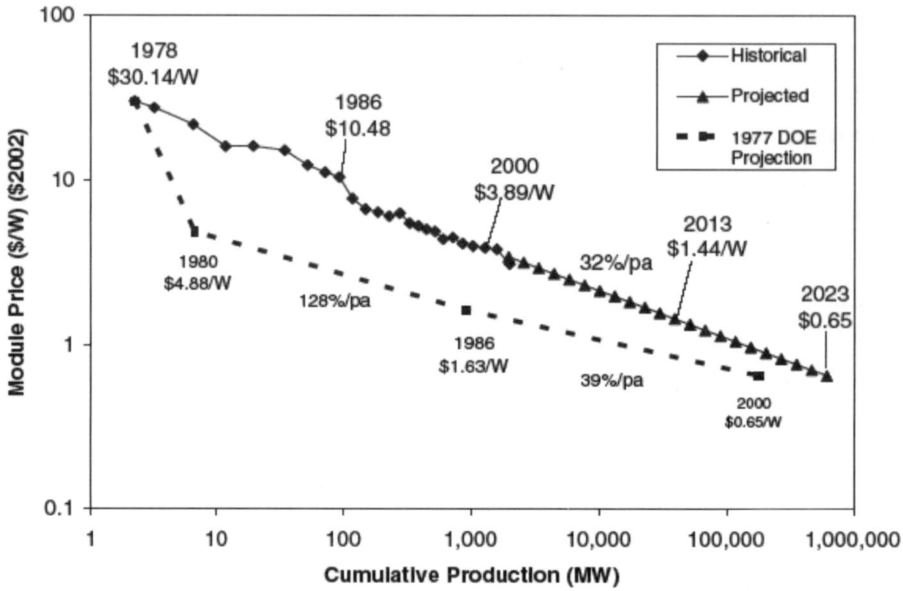

Fig. 10.8 *Extension of the historical experience curve until 2023, assuming a continual 30% annual market growth*

Extending the projection, it is seen that the module cost will touch $0.65/ watt by 2023. This cost could possibly be realised by reaching to 50 μm thick wafers and 50 μm kerf loss. Whether these figures could be reached which are considered unlikely for wafered silicon (Ref.12) is a poser to the PV industry.

The DOE (of USA) projections of 1978 are also included in the figure, which expects the module cost to reach $0.65/w by 2000 itself; a too optimistic a projection which has not happened.

Recent efforts for lowering manufacturing costs:

Stefan Rein of Fraunhofer Institute for Solar Energy Systems argues for monitoring of cell production using fast, straight forward in-line systems that can be integrated into the automated process (Ref.15). The throughput of modern production lines has risen dramatically because they handle around 2400 cells per hour, with a few individual systems even able to process up to

3600 cells. He wants to lower the cost of PV by increasing process quality in terms of increasing production efficiency, lowering rejection rates and using less expensive silicon. In such a case, not only the process but the quality of feedstock needs to be constantly monitored. The complex measurement equipment hither to used for quality evaluation have to be replaced by fast in-line measurement systems that would ensure efficient process control and quality assurance. With wafers becoming thinner, falling below 200 μm, contact less processes will be desirable.

Lock-in thermography, photoluminescence, and electroluminescence are the methods, which he suggests for fast in-line process monitoring and output control. These are good options but require much more research for implementation.

10.3.5 Status of Thin Film Technologies

As already seen, the unique advantages of TF technologies over crystalline silicon are: ease of large area manufacturing, choice of several techniques for preparation, lower cost, higher kWhr/kW output, flexible substrates, attractive appearance, and scope for innovation and improvement. Since 2000, their production volumes are around 6-8% of the total PV modules produced (Ref.16).

According to the recent study, 'PV Thin Film Industry Analysis', conducted by EuPD Research, the global market is currently shared among 116 companies working at different development levels of the five thin film technology types. In 2007, the total thin film modules produced by the top 10 manufacturing companies amount to 735MW, and the projected figure for 2008 is 1371MW (Ref.17). First Solar of USA tops the list with 308MW in 2007, and plans for an increased production of 495MW in 2008. This American CdTe market leader is planning for gigawatt dimensions. Sharp's new plant announced a production of 1 GW of thin film cells with latest triple-junction technology, while its present production of thin film a-Si tandem modules will grow to 160MW. There is an increased commercial interest in TF technologies around the world, but particularly in the USA in recent years. Ullal estimates that the worldwide thin-film production capacity will be 3.7 GW in 2010, with Japan accounting for 1300MW, USA accounting for 1200MW, and Europe and rest of the world for 1200MW (Ref. 18). Refer to chapter II in this contact.

However, there are several issues in processing and manufacturing stages to be addressed. The critical area is maintaining 'process control' and 'high yield' because of the large area fabrication running to a meter wide and a few kilometer long roll. In-line process control and diagnostics have been introduced in the manufacturing of Cu(InGa)Se_2 (Ref.19) and a-Si based modules (Ref.20). Regarding 'yield', BP Solar for a-Si tandems (Ref, 21) and Shell Solar for Cu(InGa)SeS_2 (Ref.22) have reported yields in excess of 95% from the actual manufacturing lines which is considered as a reasonable value.

Another critical issue is 'throughput'. Hegedus has discussed this issue for different thin film modules in his review (Ref.16). Rapid thermal processing (RTP), and the low-cost non-vacuum processes are suggested for future to lower the manufacturing costs of thin film modules. The combination of the two also may result in lowering the manufacturing costs, but requires more investigations. RTP has already been applied to Cu(InGa)Se_2 modules (Ref. 23) and to recrystallise a-Si films into large grain poly silicon or nano crystalline-silicon.

Incorporating thinner layers, integration of thin film modules using laser or mechanical structuring for interconnection are other methods to reduce the manufacturing costs substantially.

Several challenges for scaling up to large area, high throughput manufacturing of thin film modules are discussed by many groups (Refs. 24-27).

10.3.6 Large Scale Manufacturing of TF Modules

Large scale manufacturing leads to lower cost compared to crystalline silicon. The production volumes must be sufficiently large to offset the capital costs. However, it is not clear whether the selling prices have any relation to manufacturing costs because the manufacturers are not revealing the costs of installation, production and operation of the facilities (Ref.16). Several studies related to the costs of large scale production have been reported (Ref.28-31).The recent study at NREL (Ref.31) provides a detailed analysis of thin film module manufacturing at gigawatt scale, describing a series of module, plant and equipment design aspects leading to a-Si, CdTe and Cu(InGa)Se2 modules under $1.00/watt. Besides this study, all other studies also conclusively agree that all TF solar PV can meet the target module cost of less than $1.0/W with reasonable assumptions, when manufactured on large scale, 100MW or larger. The other factors to be considered in such large scale manufacturing are cloning of equipment, efficient material usage, frameless modules and spray-on encapsulants.

While all the industry related aspects deserve very high consideration for scaling up the production to bring down the costs, the creation of markets that demand this scale of module production deserve equal attention because for sustainability, both go hand in hand of custainability. Markets in countries where there are tremendous opportunities such as China, India and African countries have to be explored.

According to EPIA, in the next 20 years, the market frontrunner will be the off-grid applications outside the USA, Japan and Europe. Their share of the global PV market is forecast to increase to 30% by 2030 (Ref. 32). Swiss Bank Sarasin's forecast is much higher, a market volume of 18GWp in 2012. Sarasin expects that TF systems will see an average growth rate of 65% during the next five years, considerably higher than the growth rate of PV itself. The Bank further predicts that a-Si, CIGS and CdTe technologies will achieve module efficiencies of greater than 10% by 2012, with greatest prospects for a-Si technology (Ref.32).

Large scale PV power plants:

A few grid-connected large-scale plants are mentioned under Section 10.2 in this chapter. Several large scale PV power plants have been operating for more than two decades in Germany, USA, Netherlands, Spain, Italy and a few other countries.. The plants have utilized wafered silicon as well as thin film modules

The Largest Roof top solar PV of 12 MW capacity on the top of GM car manufacturing facility in Figuerueles, Spain covers about 2 million sq.ft. The project utilizes United Solar Ovonic's thin film flexible solar laminates and Satcon's Powergate Plus 500 kW inverters. The installation is expected to generate 15.1 million kWh of power annually. (www.solarcellsinfo.com/blog/archives/category/si-solar-cells).

The biggest power plants completed/operating in 2004 are listed in the International issue of Sun & Wind Energy (Ref.33).

1. The 5 MWp Leipziger Land Solar park in Espenhain, Gernay uses 33,500 Shell Solar SQ 150-LC frameless modules, and 12 Siemens Sinvert 400 kVA inverters. The yield is 950kWh/kWp.
2. The Springerville Generation Station Solar System (SGSSS) in Arizona, USA of capacity, 4.6 MWp is planned by Tucson Electric Power. The plant uses 34,980 modules of ASE, First Solar and BP Solarex, and Xantrex PV-150 inverters. The capacity is likely to be increased to 8.6 MWp.
3. Several projects with capacities up to 4MWp completed in Germany, USA, Netherlands and Spain are also given in the reference.

Hegedus describes a few large scale thin film installations under successful operation, using a-Si, CdTe and Cu(InGa)Se2 modules (Ref.16). Regarding the stability of TF modules, a-Si modules, after initial stabilization, degrade by the same amount as c-Si modules (Refs.34-36). The Cu(InGa)Se2 and CdTe modules are similarly stable, though the stability studies are less (Refs.37-40). However, their long-time stability is still a concern and investigations are essential to create more confidence.

10.4 ECONOMICS OF PV

The rapid growth of PV market and its effect on the falling of costs were seen in the earlier section. The mass-scale production and improved technologies and processes are cited as the reasons for this trend. It is to be noted that the cost reductions apply to the PV cells and modules. The costs of balance of system components are not falling as rapidly as the costs of PV modules. This has led to extensive research activities in the areas of balance of systems, and also in developing appliances that can be powered directly from the PV modules and arrays. More volumes of BIPV also reduce the costs of balance of systems.

The relative proportion of costs of each element in a PV system from 1990 to date and projected values to 2020 are shown in Fig. 10.9. Two factors are clear from this figure: the costs of cells and modules make up a substantial fraction of the total cost and the costs of PV are decreasing over time.

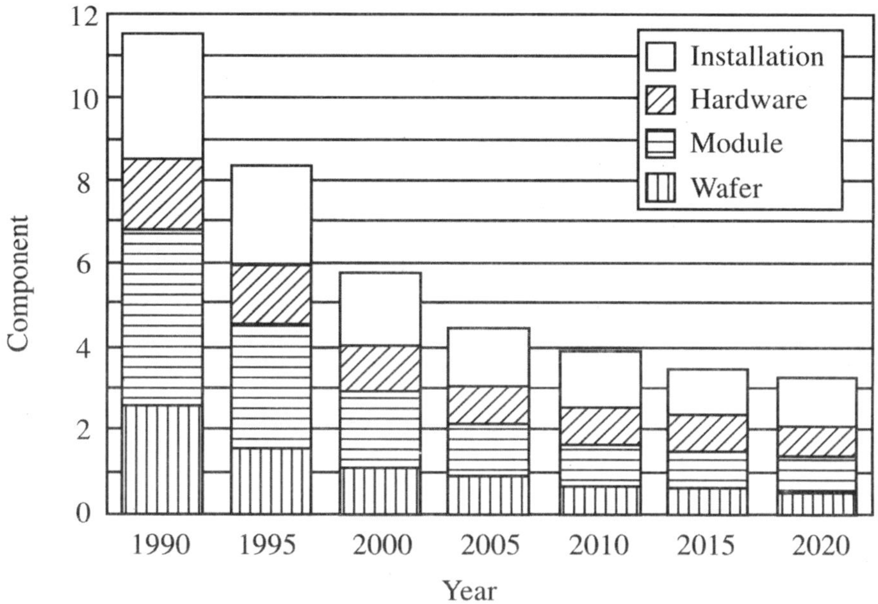

Fig. 10.5 *Component costs of PV System and Cost reduction with years*

The economic viability is assessed by using life-cycle costing, ie, the initial costs and all future costs are considered for the entire 'operational life' of the System. The typical life expectancy for different PV components offered by manufacturers are:

PV modules	-	25 to 30 yrs,
Charge controllers	-	15 yrs,
Inverters	-	5 to 15 yrs,
Batteries	-	3 to 5 yrs (solar) 5 to 10 yrs (stationery with shallow cycles)
Wiring	-	10 yrs

The period taken for analysis is normally the lifetime of the largest lived system.

Since PV is a modular system, the cost per kwh of PV power is not very much influenced by the load. The relative economics, however, are affected by the price of PV per peak watt.

The life-cycle cost analysis of a typical case is given as an example.

Typical Case: The analysis relates to the 'residential system' discussed in Annexure 8. The PV System comprises AC and DC loads, battery, invertors and BOS. The descriptions of these items are given in the Annexure 8.

Item	Cost worth (US $)	Present cost (US $)	% total LCC
1. Capital Costs :			
PV Array	3720	3720	20.8
Battery	1400	7.8	
BOS+Mounting	1635	1635	9.2
Inverters	3750	3750	21.0
Installation	1300	1300	7.3
(A) Subtotal (Equipment & Installation)	11,805	11,805	66.1
2. Operation and Maintenance :			
(B) Annual Inspection	125	1860	10.4
3. Replacement			
Battery bank 5th year	1400	1208	6.8
Battery bank 10th yr	1400	1042	5.8
Battery bank 15th yr	1400	899	5.0
Inverters			
Rebuild 10th year	2000	1488	8.3
Controller 10th year	210	156	0.9

Contd....

(C)	Subtotal			
	(Replacement Cost)	6418	4793	26.8
4. Salvage				
(D)	20% of original 20th yr	2361	609	3.3
	TOTAL Life-Cycle Cost: (A+B+C+D)	---	17,849	100.0

The cost of the ac distribution panel, generator transfer switch and battery charger is included in the BOS cost.

In summary, the relative percentages of ' net worth' costs of different items in the total Life-Cycle Cost are:

PV array: 20%, Battery: 8%, BOS: 9%, Inverter: 20%, Installation: 7%, Operation & Maintenance: 10% and Replacements: 26% (Source : www.sandia.gov/pv/sysd/ResidenceSize.html)

REFERENCES

1a. ***Solar and Wind Energy:*** *Solar home systems for rural electrification, p67(2/2004)*

b. ***B. McNelis et al:*** *2nd World Conference and Exhibition on PV Energy Conversion, Vienna(1998)*

2. ***News item:*** *Renewable Energy World, 5, no.2 (2002)*

3a. ***N.Enebish et al:*** *Renewable Energy World, 5, no.1, p83 (2002)*

b. ***Solar and Wind Energy:*** *Solar operated pumps, p64 (2/2004)*

4. ***M.Oliver et al:*** *Energy Policy, 27, p371 (1999)*

5. ***C.Nyman:*** *2nd World Conference and Exhibition on PV Energy Conversion, Vienna (1998)*

6. ***M.A.Green:*** Prog. Photovolt. Res. Appl. 14, p383 (2006)

7. ***K. Zweibel :*** The terawatt challenge for thin film PV, Tech. Report NREL/TP-520-38350, August 2005

8. EPIA-Greenpeace Solar Generation IV Report, 2007

9. PV News, May 2007

10. ***Swiss Sarasin Bank,*** Solar Energy Report 2007.

11. ***M.Rogol:*** Photon International, p102, June 2006

12. ***R.M.Swanson:*** Prog.Photovolt.Res. Appl. 14, p443 (2006)

13. ***R.M.Swanson:*** Proc.12th NREL Silicon workshop, Breckenridge,CO,USA (2002)

14. ***Zhao J et al :*** Proc.15th Intl. Photovoltaic Science & Engg. Conf., Shanghai, p310 (2005)

15. ***Solar&Wind Energy:*** In-line quality Assurance: Checking Cells during production,p126 (6/2008)

16. ***S.Hegedus :*** Prog.Photovolt.Res.Appl. 14, p393 (2006)

17. ***Solar&Wind Energy:*** PV industry – Growth across the Board p122 (4/2008)

18. ***Solar&Wind Energy:*** USA at a turning point, p99 (5/2008)

19. ***I.Rippins et al:*** Materials Research Society Symposium Procedings, Vol.865, p499 (2005)

20. ***M.Izu et al:*** Solar energy Materials & Solar cells, 78, p613 (2003)

21. ***R. Arya et al:*** Prog. Photovoltaics, 10, p69 (2002)

22. ***R.Wieting:*** Proc.29th IEEE Photovoltaic Specialist Conf., New Orleans, p478 (2002)

23. ***J. Palm et al :*** Solar Energy, 77, p757 (2004)

24. ***R.Wieting et al :*** Proc.31st IEEE Photovoltaic Spec. Conf., Orlando, p177 (2005)

25. ***T.Russell et al:*** Progress in Photovoltaics, 5, p353 (1997)

26. ***B.Rech et al :*** Proc. WCPEC, Vienna, p391 (1998)

27. ***P.Sheldon:*** Progress in Photovoltaics, 8, p77 (2000)

28. ***J.Woodcock et al:*** Proc.14th European PVSEC, Barcelona, p857 (1997)

29. ***P.Maycock:*** Proc.14th European PVSEC, Barcelona, p869, (1997)

30. ***K.Zweibel:*** Solar Energy Materials and Solar cells, 59, p1-18 (1999)

31. ***M.Keshner et al:*** NREL/SR-520-36846, 2004, available on line at www.nrel.gov/docs/fy05osti/36846.pdf

32. ***Solar&Wind Energy:*** EUPVSEC- Industry meeting in Valencia, p102 (5/2008)

33. ***Solar&Wind Energy:*** The Next Dimension, p18 (2/2004)

34. ***D.Osburn:*** Proc.American Solar Energy Society Conf. Austin (2003) {www.nrel/pv/thin_film/pn_techbased_amorphous_silicon.html}

35. ***J.Adelstein et al:*** Proc. 31st IEEE PV Specialist Conf. Orlando p1627 (2005)

36. ***A. Gregg et al :*** Proc. 31st IEEE PV Specialist Conf. Orlando, (2005)

37. ***WN. Shafarman et al:*** In Handbook of Photovoltaic Science and Engg, Luque A & Hegedus S (eds), John Wiley and Sons, UK, p567 (2003)

38. ***F.Karg et al:*** Proc.17th European PVSEC, Munich, p391, (2001)

39. ***B.Kroposki et al:*** Proc.25th IEEE PV Specialist Conf. Washington DC, p969 (1996)

40. ***First Solar application note SPR-104 TEP4.*** Tucson electric Power Springville Generating Station System Performance Report, Oct.2005 (www.firstsolar.com)

ANNEXURE - 1

MEASUREMENT OF I-V CURVE OF A SOLAR CELL

ANNEXURE - 1

MEASUREMENT OF I-V CURVE OF A SOLAR CELL

The Current-Voltage (I-V) characteristics of solar cells are measured with a solar Cell Tester and Sorter. The main features of this instrument are:

* Evaluates solar cells of up to 2.5 peak watt
* Generates I-V curve and computes and stores all important parameters such as Short circuit current (I_{sc}), Open circuit voltage(V_{oc}), Peak power(P_p), Voltage at peak power(V_p) and Current at peak power(I_p),
* To ensure highly accurate and stable measurement, a precision feedback loop for insolation control is incorporated,
* Holds temperature to the specified test value during the test
* Can measure the cell current at any preset voltage
* Grades the cells into different groups based on power output or current at any given voltage
* Effects temperature correction for the measured parameters

* Can compute the statistical parameters like mean, standard deviation and Variance for the measurements carried out over a number of cells

The System configuration and Specifications are the following:

* Sun Simulator : A xenon lamp ensures an intensity of 100 mW/sq. cm with uniform distribution throughout the test bed. A precision feedback control ensures stable illumination intensity,
* PC Based System Controller: Pentium based CPU at 166 MHz, 8 MB RAm,1 GB HDD, 1.44 MB FDD, 101 keys key board, Monochrome monitor, 80 column printer
* Electronic load : Computer controlled load enables accurate and automatic setting of cell parameters,
* High Speed Acquisition System: Consists of 16 channel AD-DA card with interface electronics to facilitate data aquisition,- Voltage measurement range, 0-1 V, and Current meausrement range, 0-5A.
* Water cooled Test Bed with vacuum hold and spring loaded gold-coated test probes,

Temperature Measurement Accuracy	0.5^0.C,
Temperature control accuracy	1.0^0.C,
Temperature range	$0\text{-}50^0$C

* Application Software Package : Menu driven user friendly applications software package enables setting the parameter for grading the cells; Controls data aquisition; Performs data processing, formatting and storing; Complies I-V characteristic; Computes both performance and statistical parameters for evaluation of performance of cells under test.

A typical I-V curve and voltage-power curve obtained for a typical solar cell with this instrument is given below.

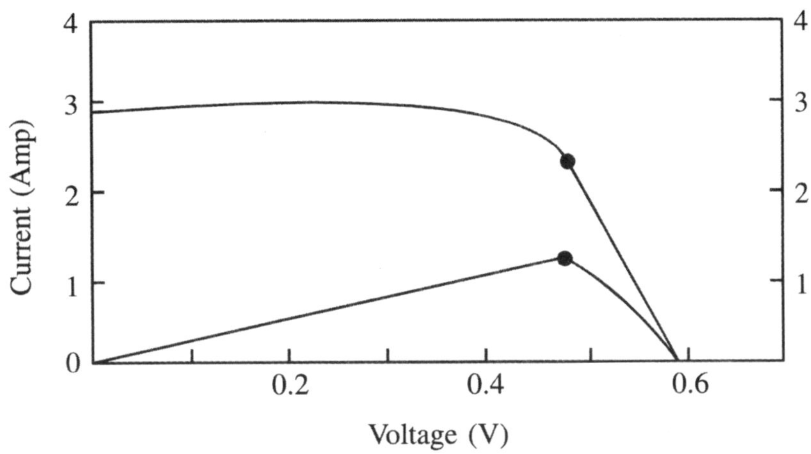

***Fig. A.2** Measured IV curve of a typical Solar Cell*

The test is performed at 25 deg.C with an insolation of 100 mw/sq.cm. on a solar cell of area 100 sq.cm. The data obtained are:

V_{oc} = 589 mV, I_{sc} = 2.89 amp

V_{mpp} = 481 mV I_{mpp} = 2.53 amp

P_{mpp} = 1.22W FF = 71.7%

and Efficiency = 12.2%.

ANNEXURE - 2

SOLAR RADIATION MEASURING INSTRUMENTS

ANNEXURE - 2

SOLAR RADIATION MEASURING INSTRUMENTS

The principles of working of the instruments used to measure the global (total), diffuse (sky) and direct radiations are briefly outlined:

PYRANOMETER

This instrument is used to measure the 'global' and 'diffuse' solar radiation. It consists of a double dome made of quartz or glass which freely transmits solar radiation falling on it without any interference. There is a small 'black' surface at the centre surrounded by a 'white' surface in the form of a ring. These two surfaces act as sensors,- the former absorbing the radiation and the latter, reflecting the radiation. The temperature difference between the two surfaces when exposed to solar radiation is measured using a thermopile. The output is generally in milli-volts which can be handled by a variety of data recording systems. Eppley Pyranometer, an extensively used instrument works on this principle. A schematic diagram showing different components is shown in Fig.B.1.

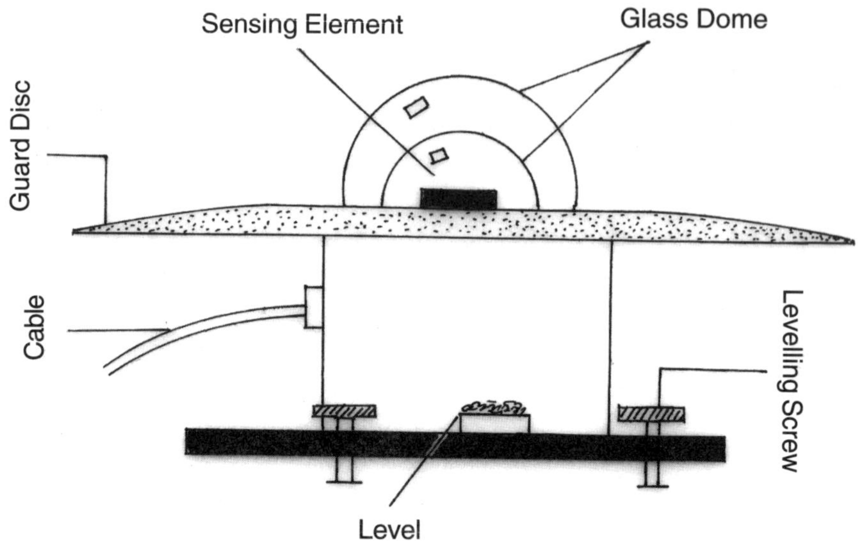

Fig. B.1 *Eppley Pyranometer*

This pyranometer with 180 deg. field of view measures 'global' radiation. If a shadow band is arranged to the instrument, it can block the 'direct' beam radiation, and the instrument measures only the 'diffuse' radiation. Some correction is needed to the measured value and the correction factors are provided by the manufacturer.

Rotating Shadow band Radiometer is another instrument which measures both global and diffuse radiation. The principle of operation is the same; a motor moves the shadow band to block the sensor from 'direct' beam solar radiation, allowing the diffuse radiation measurement. To measure the global solar radiation, the motor moves the shadow band to un-shade the sensing element.

In some measurements, the shadow band is replaced by a shadow disc. This circular disc is about 10 cm in diameter and is positioned to rotate at a distance of about half to one meter in front of the pyranometer to shade only the pyranometer. The correction aspect is considered to be less in this type compared to the shadowed type.

PYRHELIOMETER

This instrument is used to measure 'direct' solar radiation. It has a narrow field of view and senses only the direct beam coming from the sun. Angstrom Pyrheliometer is one of the earliest instruments; it consists of a sensing element which gets heated up when exposed to solar radiation. The heating effect is matched by an electrically heated 'shaded' element, and the electrical heating is measured by any of the conventional standard methods. The schematic of pyrheliometer is shown in Fig.B.2.

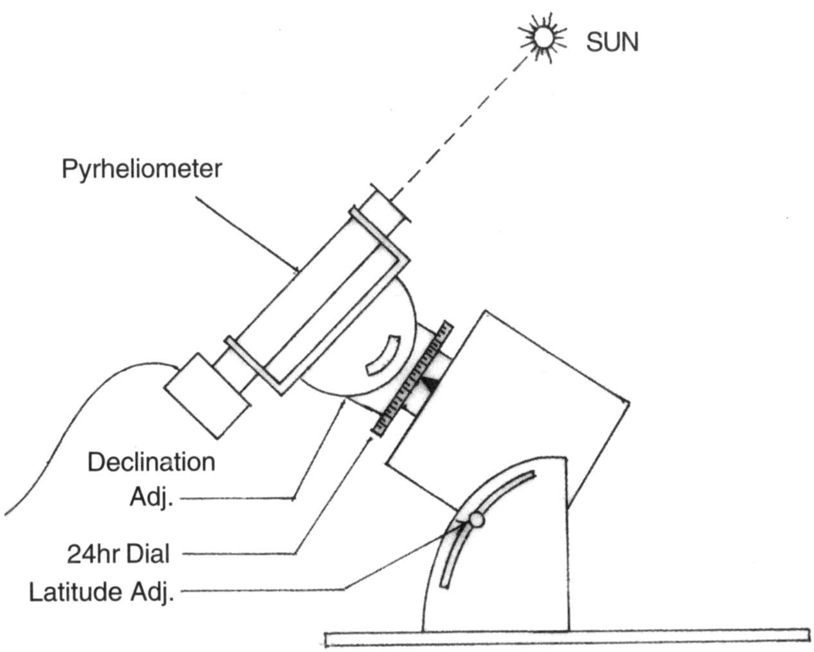

Fig. B.2 *Schematic of Pyrheliometer*

Abbot Water flow calorimeter is another instrument; this contains a cavity that absorbs solar radiation falling on it. As a result, the temperature of the circulating cooling water rises and this rise is proportional to the incident solar irradiance.

Initially, the measurements made with different instruments disagreed to an extent of 2.5 to 6%. The International Pyrheliometric scale was developed to apply corrections to the measurements.

PYRANOMETER AND GLOBAL RADIATION INTEGRATOR

This instrument measures the instantaneous solar radiation at any given location and displays on a LCD digital panel meter. It also provides the total insolation recorded over a given period of time through an electronic counter.(Plate B.3)

The specifications of the instrument are:

* Instantaneous insolation display(16 digit dot matrix alpha-numeric LCD display)
* Resolution- 1 watt/sq. m.
* Energy display with 0.01 kwh resolution
* Uses calibrated reference cell
* Operates on 230V, 50Hz AC or on 12V built-in battery for field operation
* Constant current charger for 12V battery
* Battery low indication
* Battery overcharge indication

ANNEXURE - 3

PV MODULE FABRICATION

ANNEXURE - 3

PV MODULE FABRICATION

The processing steps for fabrication of solar PV modules are briefly outlined here.

(i) Inspection of solar cells

Solar cells are available in circular, square and pseudo-square shapes (Fig. C.1). These cells are examined visually for any damage and for dimensional non-conformance as per the specifications. The cell shall have 'front n-contact' with grids and bus bar and the 'back p-contact' completely covered. Both the contacts shall have pre-coated solder. Solder coating on the solar cell contacts should be free from defects such as excess or overflowed solder and flux on the cell's active area. If any physical damage is suspected, it shall be examined under a magnified lens/microscope to know the extent of damage.

The solar cells are liable for rejection if any of the following defects are noticed:

* deep scratch on the surface or crack in the cell,
* twin plane in the cell,

* any gap/corrosion/inclusions causing discontinuity in n-contact bus bar or grid lines or back contact, and

* any loose grids or contacts.

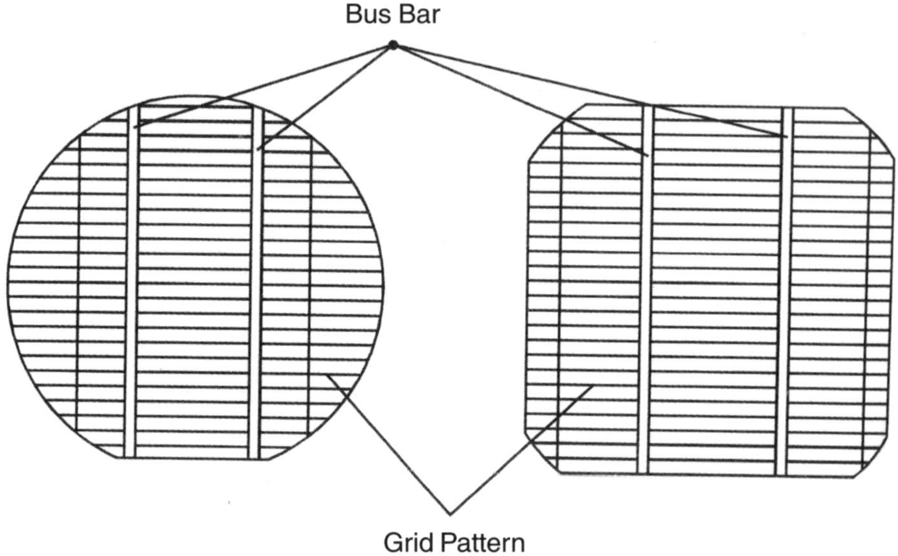

Fig. C.1 *Solar Cells showing Grid-pattern and Bus Bars*

(ii) Solar Cell Evaluation

The best way to evaluate the electrical performance of a cell is to study current-voltage behaviour under steady state simulated sun light. The measurement of I-V curves are given in Annexure 1.

From this data, the cells are sorted according to their current rating. The cells that are interconnected in series shall have the same current rating. The cells that go into a module also shall have the same current rating within a variation of about 10%.

(iii) Preparation of Cell Interconnects and Bus Bars

This step is very important. Oxygen-free copper foil of length about 9 to 10 cms, width around 0.25 cm and thickness less than 0.01 cm is used for interconnects.

For bus bar, oxygen-free copper ribbon of length about 15 cms, width around 0.5 cm and thickness around 0.01 to 0.02 cm is used .The shapes of the 'interconnect' and 'bus bar' are shown in Fig. C.2. All the interconnects and bus bars are coated with a film of solder

Solder material (Sn-Pb alloy) on both sides by electroplating. The solder thickness on each side of interconnects and bus bar shall be around 0.01 mm. All interconnects are visually examined for any defects like scratches, bends, cuts, dents, blistered solder coating and corrosion. Those with none of these defects are selected and stored in sealed polythene covers.

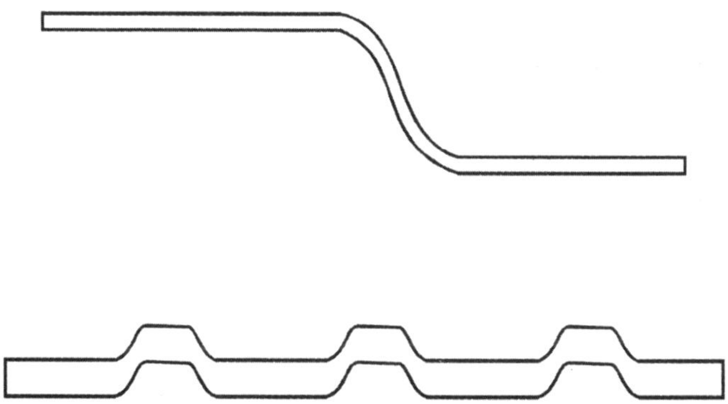

Fig. C.2 *Shapes of Interconnect and Bus Bar*

(iv) Front Contact Tabbing

The process sequence of the front contact tabbing of solar cells is as follows:

Both surfaces of the solar cell bus bars and interconnects are properly cleaned using ethanol. The cells and the inter connects are then fixed in a soldering jig, carefully aligning the interconnects. Make sure hat the interconnect tabs are making contact with bus bar grids on the cell. The solder material is Sn-Pb-Ag alloy in the form of a wire with cored RMA flux. It shall be tested before use for composition, spreading and wetting. The solder flux is a rosin (mildly activated liquid) and shall have about 35 to 40% of solid content. The solder flux is also tested for its flow, consistency, spreading and dryness.

The temperature of the soldering station is set at 270 ± 10°C, and the tip of the soldering iron is cleaned with wet sponge to remove any oxidized matter. The solder material is applied from the opposite side of the soldering tip and allow it to flow completely around the interconnect and along its length on the cell. The solder joints of front contacts are cleaned using MEK followed by smooth ethanol wiping. The measured bond strength must be greater than 500 gms when tested in 90 degree peel mode.

Before accepting, it is to be ensured that

* there shall be proper amount and distribution of solder on the joint which shall appear smooth and bright
* there shall be no pinholes or voids in the joint,
* the cells, contacts, interconnects and solder joints are not damaged or fractured during soldering, and
* no trapping of excess flux and other contaminants in the solder joint

If any of these are not complied with, then reworking/repairing is necessary.

(v) Interconnection and Bussing of Solar Cell Circuit

The cells which go into the module are interconnected in series and in parallel to get the required power from the module at the specified operating voltage. The cells with front interconnect tabs are visually examined for any defects. The interconnects and back contact surface of the cell are cleaned and the cell is aligned in the fixture made of polyethylene. The fixture contains grooves having the exact cell dimensions and also provision to accomadate interconnects as per the module layout. The typical module layout is shown in Fig. C.3. The depth of the groove has to match the cell thickness and front solder contact. Cells are aligned in the grooves of the fixture with upside down and the free end of the interconnect is placed in the backside of the next cell. Check to ensure that the interconnects are touching the cell surface uniformly and also they make proper contact with back surface of the cell. Soldering is performed as outlined.

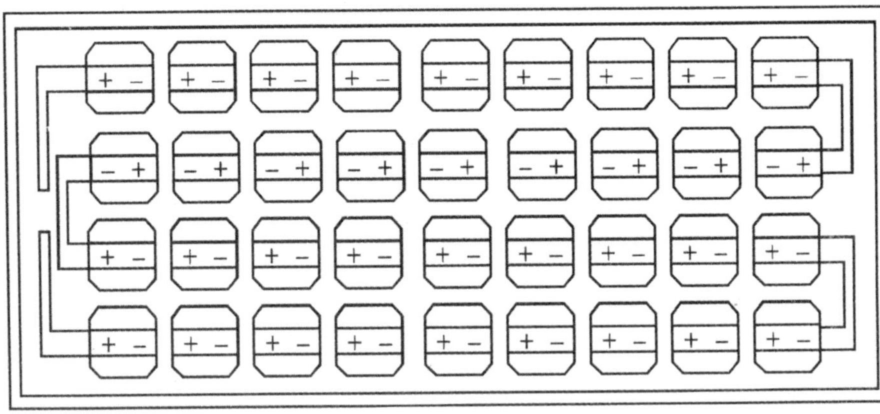

Fig. C.3 *Module Layout*

The solder joints are cleaned with MEK followed by ethanol wiping. The solder bonds of interconnects are examined for any defects that occur during soldering. It is to be ensured that solar cells, contacts and inter- connects and solder joints are not damaged during soldering.

The test for solder bond strength is then performed. The bond strength is measured by clamping the cell with solder interconnects in the fixture and pulling the tab at 90 deg. to the cell surface at a pull rate of greater than 30 mm per minute.

The interconnected solar cell circuit is lifted from the fixture and reversed with the aid of module reversing fixture. The interconnects and bus bar are cleaned , and keeping the module in the fixture, the bus bars are aligned. Ensuring that the interconnect ends are making proper contact with bus bar, soldering is done as outlined above. The solder joints are cleaned using MEK followed by ethanol wiping; and inspected for any soldering defects in the solder bond of interconnects. The bond strength measured must be greater than 500 gms in 90 degree peel mode.

(vi) Module Cleaning and Inspection

The front and back surfaces of the solar modules are cleaned before inspection for their worthiness.

The module is placed on a table covered with FEP film and the front surface of the module is cleaned with a tissue paper wetted with ethanol. The tissue paper is moved during cleaning in one direction only. The module is reversed using module reversing fixture and the back surface of the module is cleaned as detailed already.

During cleaning, ensure that either the cells or interconnects or solder joints are not damaged and also that no tissue fibre is left on cell surface.

The modules are examined for any defects, such as cell cracks, deformed interconnects, loosened contacts and solder joint defects. If any of the defects are present, they shall be repaired or replaced.

(vii) Lamination of Modules

The lamination (or encapsulation) is the most important step in the module fabrication. It involves preparation of lamination materials, assembly (or stacking) of lamination materials and vacuum lamination of the assembly.

The materials are glass plate, EVA film, fibre glass mat and back cover material. The glass plate which is the' front cover' material is a low iron, tempered soda-lime glass sheet of 3 to 4 mm thick capable of providing mechanical durability, weatherbility, and weather-resistant bonding to the pottant. It shall have hemispherical transmittance greater than 91% and shall be stable in the terrestrial environment for the module life of more than 20 years. A plate cut to the dimensions of the module area is cleaned on both sides with isopropyl alcohol using a tissue paper. EVA film, fibre glass mat and back cover are cut to the size of about 5 cms in excess to module dimensions. EVA sheet of about 0.5 mm thick is used as a pottant which encases solar cells. EVA has high optical transmittance and free from odour and toxicity. Its surfaces shall have a suitable pre-coated primer, and its glass-transition temperature must be less than -40^{0}C. It shall be compatible with the state-of-art module fabrication process by vacuum lamination. The non-woven glass mat is used as air-release porous spacer and has a thickness of 5 mils. It is electrically an insulator and mechanically non-compressible. It shall be porous for in-plane air flow during fabrication.

The back cover is a composite layer of tedler/polyester/tedler or tedler/aluminium/tedler combination. The tedler layer is 2 mil thick and so also the aluminium foil. This has a low surface solar absorptivity and high black surface emissivity. It serves as a backside weather barrier. All the sheets are cleaned with ethanol using tissue paper. Gaskets, silicone adhesive and mounting frames are the other materials required for framing after lamination.

Fig. C.4 shows a typical sequence of the assembly of the materials. The FEP release film, the cleaned glass plate, EVA sheet, the solar cell circuit with active side of the solar cells face down, fibre glass mat, EVA sheet, the back cover and a FEP release film are stacked in that order. Using masking tapes, the edges of the assembly are sealed firmly at a few places.

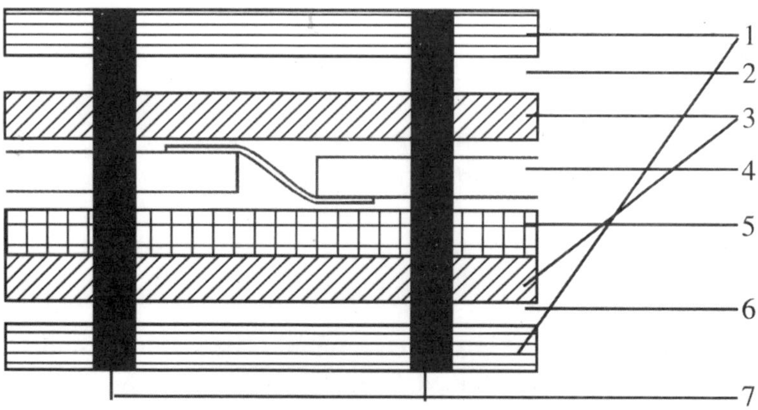

1. FEP Release Film
2. Front Cover (Glass Plate)
3. EVA Sheet
4. Solar Cells
5. Fiber Glass Mat
6. Back Cover
7. Masking Tape

Fig. C.4 *Stacking of Module Materials for Encapsulation*

The next step is vacuum lamination process using a Vacuum Laminator .It has provision for precise control of vacuum and temperature. Its design ensures that even during power failure in the lamination process the laminator materials are not exposed to environment. The assembled module is placed with the glass plate side down in the laminator, and without disturbing the assembly, the top cover of the laminator is closed. The entire assembly is evacuated to the required vacuum of 10(-3) torr. At this stage, the heating system switches on automatically and the temperature reaches the preset value of 100-120^{0}C. Under these temperature and vacuum conditions, the assembly is cured for 12 to 15 min, in the laminator. It is then cooled to room temperature and the assembly is taken out after breaking the vacuum. The masking tape is peeled off and the FEP films are removed. The laminated module assembly is cured at about 140-150^{0}C., for about an hour in automatic temperature controlled oven. The edges of the module assembly are trimmed with a blade after removing from the oven. A quick wiping of the module assembly with an ethanol soaked tissue paper is done after making sure that no damage is done to the module assembly during the vacuum lamination process.

(viii) Framing of Modules

All the laminated module stacks are framed for easy handling and transporting. The frame also functions as support channel for installation. The cross-sectional view of the 'framed' module is shown in Fig. C.5,along with the edge seal, gasket and the frame. Extruded anodized aluminium channels which have the right characteristics are used as module frames.

The cross-sectional view of aluminium channels is given in Fig.C.6. The channels are cut to the dimensions of the standard module size. A U-shaped elastomeric gasket is fixed in the aluminium channels. The gasket shall be unplasticised, extrudable and weather-stable in the terrestrial environment. It must be capable of accomadating module expansion and contraction and shall possess good compression-set-recovery property. The edge-trimmed module is carefully inserted into the groove of the U-shaped gasket on all the sides and firmly bolted.

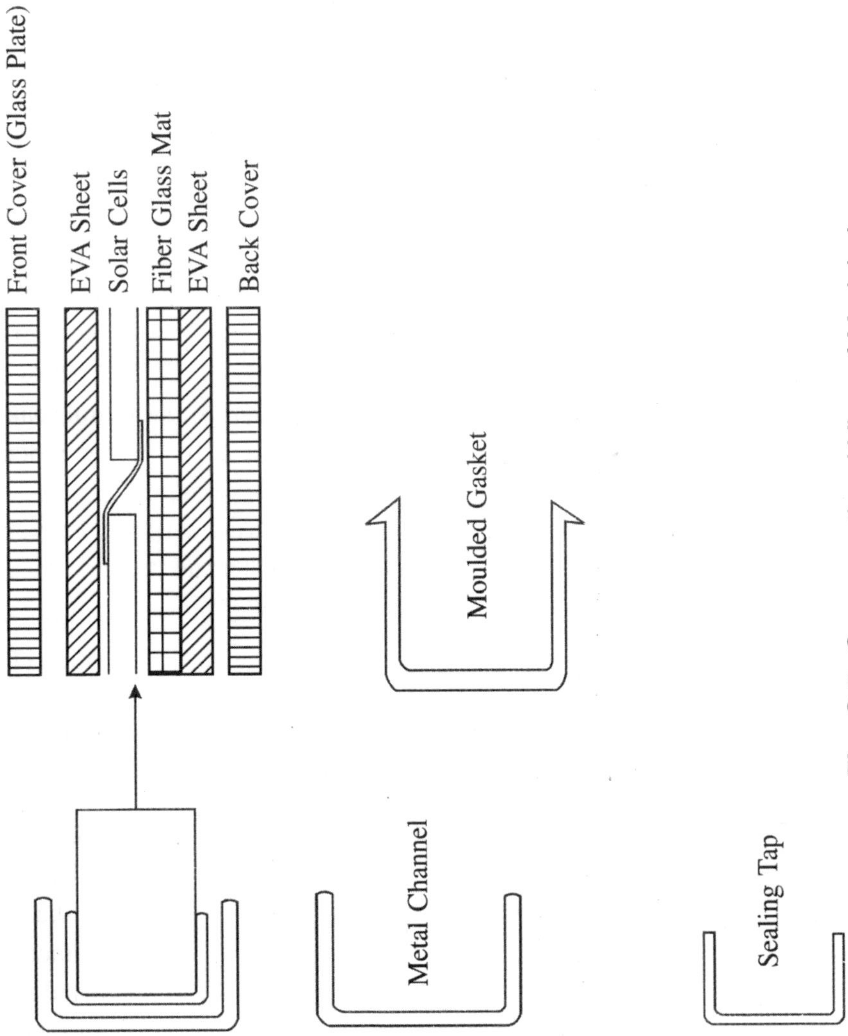

Fig. C.5 *Cross-sectional View of Module Layers*

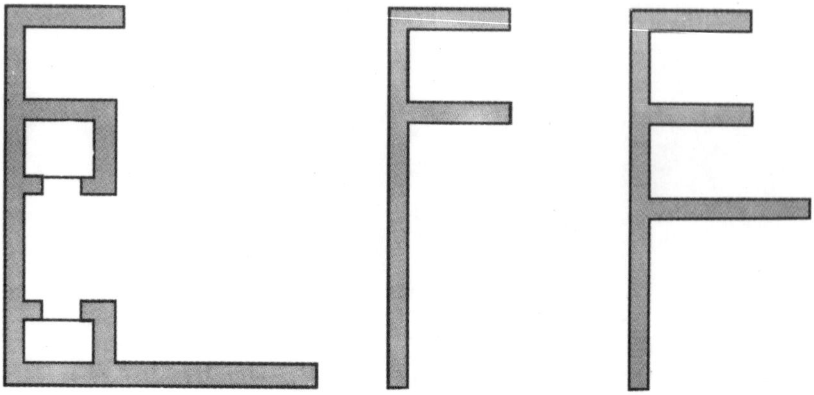

(a) Length Channel (b) Width Channel (c) Width Channel to fix terminal box

***Fig. C.6** Cross-sectional View of Aluminium Channels*

The electrical terminals of the module are connected to the terminal box which is then closed. The edge seal silicone adhesive is finally applied for atmospheric isolation of the module and cured at room temperature. The adhesive should be vulcanised at room temperature with fast curing and be chemically inert and resistant to mechanical fracture. It should be water and vapour resistant and characterised by restricted flow, creep and spread.

(ix) Module Inspection

The modules are thoroughly examined for any defects before being tested for insulation and performance. The defects to be looked for are:

* broken or cracked front cover glass plate
* chipped or broken or displaced solar cells
* broken or displaced interconnects
* solar cells touching one another or the frame
* voids, bubbles or delamination of the encapsulantss
* decoloration of the module components

* failure of adhesive bonds
* contact corrosion
* faulty terminations
* defects in the back surface

The inspection is carried out visually under bright light, not less than 1000 lux.

The modules found to be free from any of the above defects are accepted and tested for their insulation and performance. The insulation test and the Performance test are outlined in Chapter III.

MODULE ASSEMBLY LAYOUT

A typical module assembly layout is shown in Fig. C.7. Some of the tasks are performed with automated equipment to achieve high product yield. A solar cell tabbing machine is used to load cells from stacks and to align, to solder tabs to them and to unload them into stacks. The tabbed cells are then arranged on alignment boards of module-size with the tabs extending over the adjacent cells. The interconnect of the cells is done subsequently. After circuit final assembly, the completed circuit is inspected and forwarded to the lay-up station.

A vacuum pick-up tool for transferring the module circuit for the lamination process and a roll rack for dispensing sheet materials are used. The ‘completed’ stack is then loaded into a vacuum laminator which has the provision to control temperature and vacuum in a correct sequence. The modules are then cured in batches in a convection oven which has automatic temperature control. The final tasks of the module assembly such as fixing of output leads, edge seal and frame attachment and terminal box are performed using different tools and fixtures.

The performance test is done using a sun simulator. The tested modules are finally inspected before they are packed and stored.

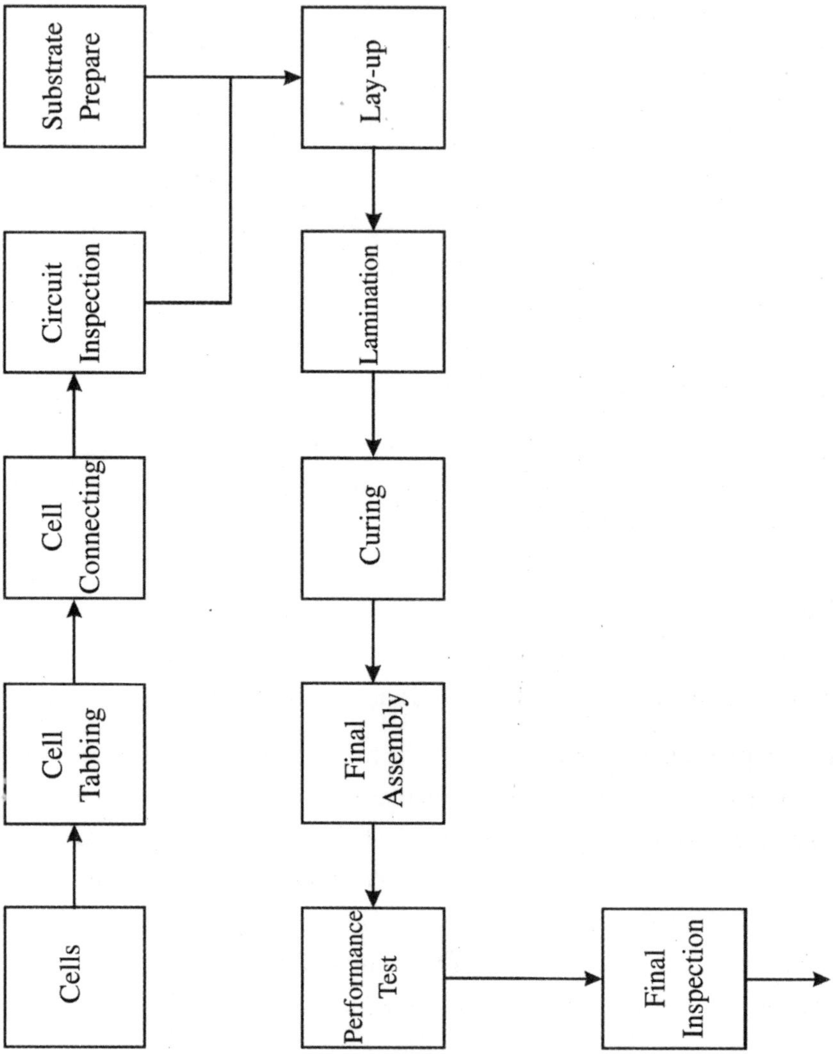

Fig. C.7 *Module Assembly Layout*

DEFECTS IN SOLDERING

Soldering is an important job in the module fabrication. The details of the solder material and the solder flux are already given. The defects generally occur during soldering if proper care is not taken. The common defects noticed are:

* cold solder connection wherein the solder is not properly flowed and wetted the interconnected surfaces
* disturbed solder connection in which stress lines and a granular zone appear
* insufficient solder connection wherein the width of the fillet is disturbed at some point
* excessive solder connection which distorts the outline of the interconnect
* non-wetting, in which the solder fails to form an inter-metallic bond with the base metal due to contaminants or insufficient flux or oxide layer obstructing the insufficiently heated molten solder/ base metal interface
* de-wetting,- a kind of surface condition wherein the solder on the metal surface irregularly collects into islands leaving a thin solder film on the surface
* pin holes which are small holes on the surface of solder connection growing into bigger ones within the connection
* voids and blow holes partly covering the soldered surface
* pit,- a surface defect whose inner surface is visible and covers more than 5% of the soldered surface
* impure connection wherein inclusions such as foreign matter, dust, flux residues, oxide layers etc get embedded in the solder surface, giving a dull non-metallic appearance *overheated connection in which, due to high temperatures during soldering, the solder appears dull or crystalline or porous

* solder splatter and projections, wherein solder balls or points or spikes or protrusions appearing on the cell surface or on the solidified solder.

Any of these defects present in the solder joints are unacceptable and reworking either by reheat method and/or re-solder method is done to correct the faulty soldering joints.

SUN SIMULATOR

The features and specifications of a Class B type Sun Simulator used for testing the electrical behaviour of solar cells and modules are briefly mentioned:

i. *it is capable of producing the standard irradiance of 1000 watts/sq. meter at the test plane with facility to increase or decrease the irradiance level as required,*

ii. *its spectral distribution matches with that of standard sunlight such that the short circuit current errors due to the possible spectral mismatch between the reference and test cells/modules are limited to 1%,*

iii. *the spectral response mismatch will be between 0.4 to 1.4 microns,*

iv. *the irradiance in the test plane as measured by a cell detector shall be uniform over the entire area of the test cell/module to within 5% of the average value,*

v. *the angle subtended by the apparent source of the simulator at any point on the test cell shall be less than 30 deg.*

The pulsed xenon/quartz halogen sun simulator which closely matches the solar spectrum meets the above specifications. The simulator needs to be checked periodically for its performance.

ANNEXURE - 4

MEASUREMENT OF IV BEHAVIOUR OF A PV MODULE

ANNEXURE - 4

MEASUREMENT OF I-V BEHAVIOUR OF A PV MODULE

The study of I-V behaviour of a PV module is measured using a MODULE TESTER with a Sun Simulator (Plate D.1). The main features of the Tester are:

* Generates I-V curve; computes and stores all the important parameters like Short-circuit current, I_{sc}, Open-circuit voltage, V_{oc}, Peak power, P_p, Voltage Peak power, V_p, Current Peak power, I_p, Efficiency and Fill Factor.
* Computes statistical parameters for the measurements carried out over a number of modules like Mean, Standard deviation and Variance.
* Effects temperature correction for the measured parameters.

The System Configuration and the Specifications are:

* Sun Simulator- with matrix of Quartz Halogen lamp/ pulsed Xenon lamp to give an intensity of 1000 watts/sq.m, with uniform distribution over a test bed area of 130 cm x 80 cm to test modules of range up to 100 Wp.

* Electronic Load Voltage Range : 0-35 V
  Current Range : 0-7.5 A

Measurement of voltage and current at increments of 10 mA of current.

* PC Based System Controller- Pentium based CPU at 166 MHz, 8 MB RAM, 1 GB HDD, 1.44 MB FDD, 101 keys keyboard, colour monitor and 132 column printer.
* High Speed Acquisition System- 16 channel ADC with 2 channel DAC on board timer and DIO
* Application Software- Menu driven, user friendly application software package for data acquisition, processing storage and graphical display of I-V curve, computes both performance and statistical parameters. Enables temperature correction for the measured parameters.

The typical I-V characteristic of a PV module determined with this Tester at a temperature of 28 deg.C and with an insolation of 1000 watts/sq.m is given below. The measured parameters are:

V_{oc} = 20.85V,

I_{sc} = 5.08A,

V_{mp} = 16.35V,

I_{mp} = 4.48A,

P_{mp} = 73.25W,

FF = 69.16%,

and Efficiency = 12.5%.

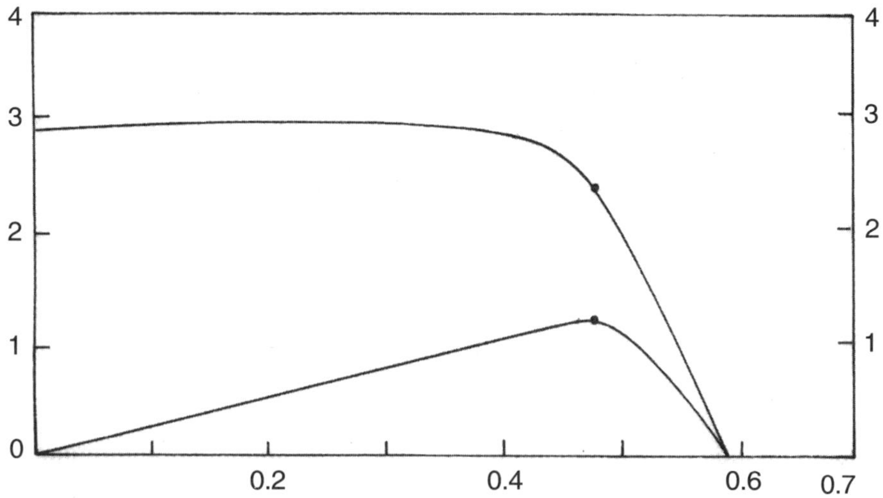

Fig. D.2 *Measured IV Characteristic of a Typical Module*

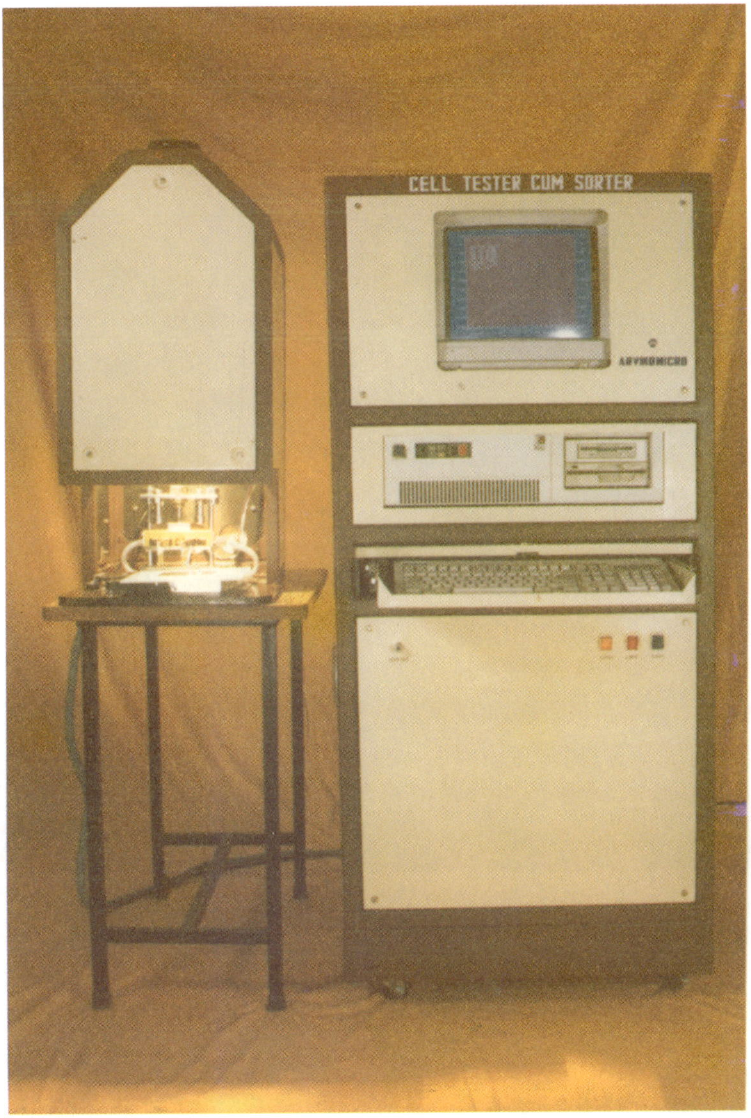

Plate 1.1 *Cell Testor-Cum-Sorter*

(by courtesy Aravind Micro., India)

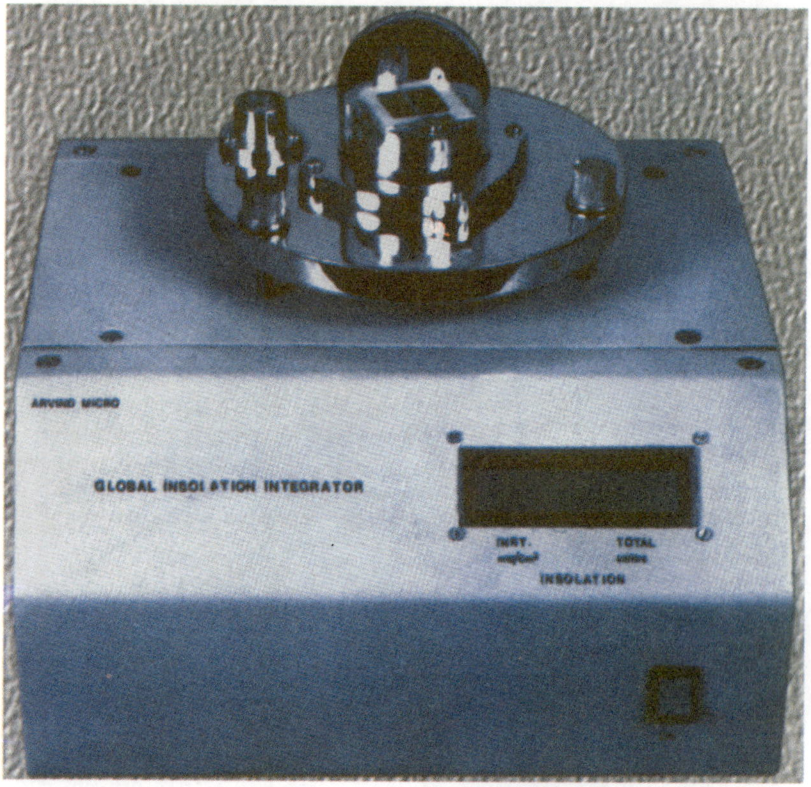

Plate 2.3 *Global Insulation Integrator*

(by courtesy Aravind Micro., India)

(a) Sun Simulator (b) Measuring Equipment

Plate 4.1 *Module Tester with Sun Simulator*

(by courtesy Aravind Micro., India)

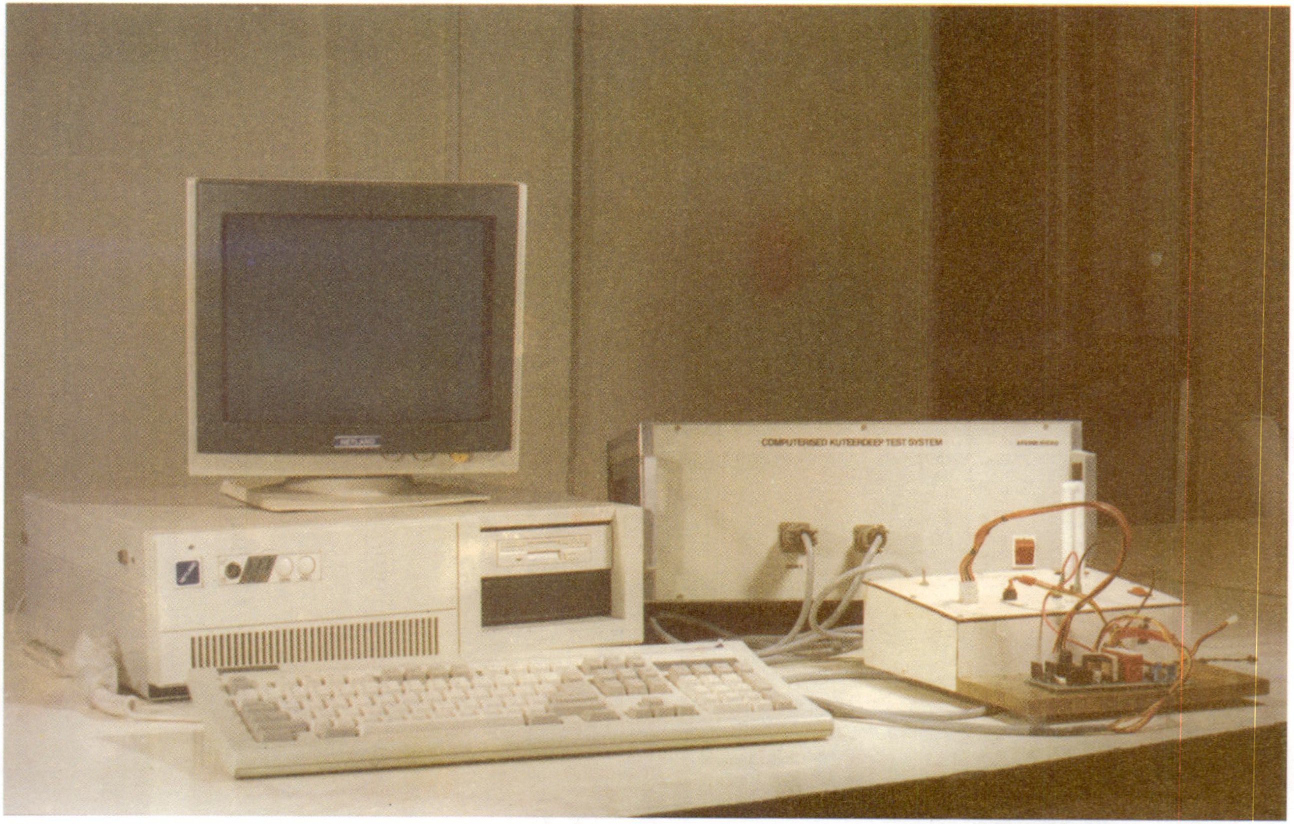

Plate: *Solar Lantern Test System*

(by courtesy Aravind Micro., India)

ANNEXURE - 5

FORMAT FOR ESTIMATING AVERAGE DAILY LOAD

ANNEXURE - 5

FORMAT FOR ESTIMATING AVERAGE DAILY LOAD

All the AC and DC appliances in the house are listed along with their wattages. The hours of usage per day must be noted to work out the 'average daily load'.

AC/DC Load Description	Qty	Wattage	Daily Duty Cycle (hrs/day)	Weekly Duty Cycle (days/wk)	Load (Whrs/day)
1. Lighting:					
Flourescent tube	....	....	....	....	....
Incandescent bulb	....	....	....	....	....
CF lamp	....	....	....	....	....
2. Fan	....	....	....	....	....
3. Television	....	....	....	....	....
4. Refrigerator	....	....	....	....	....

5. Radio ....
6. Computer ....
7. Blender/Mixer ....
8. Microwave oven ....
9. Washing machine ....
10. Vacuum cleaner ....
11. Water pump ....
12. Others: ....

....

....

....

....

(A) Total AC average daily load: ______________________

(B) Inverter Loss allowance (15%):______________________

(C) Total AC load = (A) + (B): ______________________

(D) DC Average Daily load: ______________________

Total of (C) + (D): ______________________

ANNEXURE - 6

FORMAT FOR PV SYSTEM SIZING

ANNEXURE - 6

FORMAT FOR PV SYSTEM SIZING

1. Site Data: Location ....
   Latitude ....
2. Module Specifications:
   Peak Watts ....
   Nominal voltage.........
   Amps at typical load........
3. Array Angles:
   SUMMER Array angle,(Latitude - 15 deg.) =
   (June, July, Aug.)
   FALL Array angle, (Latitude) =
   (Sept, Oct, Nov)
   WINTER Array angle,(Latitude + 15 deg.) =
   (Dec, Jan, Feb.)
   SPRING Array angle,(Latitude) =
   (March, April, May)
   FIXED Array angle, (latitude + 15 deg.) =
   (non-adjustable)

4. Usable Output of Module (UOM) :
   (Nominal Module Voltage) x
   (Module Amps at typical load) =
5. Average Daily Peak Sunhours (ADPS hrs):
   (taken from Annexure 5)
   SUMMER Ave. Daily Peak Sun hrs =
   FALL Ave. Daily Peak Sun hrs =
   WINTER Ave. Daily Peak Sun hrs =
   SPRING Ave. Daily Peak Sun hrs =
   FIXED Ave. Daily Peak Sun hrs =
   (Fall ADPS + Spring ADPS)/2 =
6. Average Daily Usable Module Output (ADUM output) =
   SUMMER Ave Daily Usable Mod. output = (Summer ADPS hrs × UOM)
   FALL Ave Daily Usable Mod. Output = (Fall ADPS hrs × UOM)
   WINTER Ave Daily Usable Mod. output = (Winter ADPS hrs × UOM)
   SPRING Ave Daily Usable Mod. output = (Spring ADPS hrs × UOM)
   FIXED Ave Daily usable Mod. output = (Fixed ADPS hrs × UOM)
7. ARRAY SIZING :
   (Number of Modules) = (Average Daily load)/(UOM)
   = ____________
8. BATTERY BANK SIZING:

The following steps may be carried out to workout the AH capacity of the 6 or 12 volts battery (which is the output voltage of the System):

i. (Average Daily Usable Mod.output)/(nominal voltage of the Mod.)................................(a)
ii. Multiply (a) by number of days of bad weather ....(b)
iii. Divide (b) by 0.7 to maintain a 30% reserve in the battery after a deep discharge ....(c)
iv. Battery bank size = 6 or 12 volts,'(c)' AH.

ANNEXURE - 7

SIZING EXAMPLE FOR PV RESIDENTIAL SYSTEM

ANNEXURE - 7

SIZING EXAMPLE FOR PV RESIDENTIAL SYSTEM

The sizing of a Residential PV System to meet the DC/AC loads of a house is discussed here. The house is located in Fort Collins, Colarado state in USA. The Latitude of the place is 40 Deg.N. and the temperatures range from –30 to +30 deg.C. It is situated in mountains and the maximum wind speed is 15 m/sec. (Source: www.sandia.gov/pv/sysd/ResidenceSize.html)

INSTALLATION

The PV array was ground mounted on a set of concrete poles about 100 ft from the residence, allowing the array to face true south and avoiding the high wind hazard. Aluminium conductors are used because of the long wire run to reduce the cost.

The house circuits were already divided into two service panels,- one panel containing the dc circuits and the other, ac circuits. The dc panel was served by the battery bank through the controller and a current limiting fuse was used in the positive lead of the battery bank. The inverter was protected by a fused safety switch, enabling it to be isolated from the battery bank for maintenance.

Estimation of Load:

Load	Qty	Load current (amp)	Load voltage (volt)	Load power (watt)	Daily duty cycle	Weekly duty cycle	Power conv.eff	System voltage (volts)	Load (amp-hrs)
					(hrs/day)	(Days/wk)			
D.C.LOADS:									
DC lights	5	1.0	24	120	4.0	7/7	1.0	24	20.0
Incand. Lights	2	2.1	24	100.8	1.0	7/7	1.0	24	4.2
A.C.LOADS:									
Well pump	1	12.0	240	2880	1.0	1/7	0.85	24	20.2
Washing Machine	1	8.4	120	1008	0.7	1/7	0.85	24	4.9
Vacuum Cleaner	1	2.5	120	300	0.25	7/7	0.85	24	3.7
T.V.	1	1.0	120	120	1.0	7/7	0.85	24	5.9
Toaster	1	9.2	120	1104	0.15	7/7	0.85	24	8.1
Microwave Oven	1	50	120	600	0.20	7/7	0.85	24	5.9
Total:				6232.8					72.9

Fused safety switches isolated the controller from the battery and the array. The negative side of the dc system and the neutral conductor of the ac system were grounded. All conductors not grounded were protected by either circuit breakers or fuses. Surge arrestors were included to suppress the transients induced by lightening. The low state-of- charge of the battery system was seen by a visual indicator.

Now,

(Total load power)/(Nominal system voltage) = Peak current

i.e., 6232.8 / 24 = 259.7 amps.

(Total amp-hr load)/(wire effi. Factor)/(Battery effi. Factor)

= corrected Amp-Hr load

i.e., 72.9/0.98/0.9 = 82.7 amp-hrs/day.

Since the ac loads are larger, the nominal system voltage is selected as 24 volts to be compatible with the inverter input voltage.

To calculate the Design current and Array tilt, the 'Peak Sun Hrs/Day' for the twelve months of the year are taken from Solar Data for three tilts,- latitude-15, latitude, and latitude+15 for the site of the installation(Fort Collins, CO). and tabulated as in Table 2.

From the above table, select the largest design current and corresponding peak sun from each tilt and enter below.

Latitude–15		Latitude		Latitude+15	
Peak Sun (hrs/day)	Design current (amp)	Peak Sun (hrs/day)	Design current (amp)	Peak Sun (hrs/day)	Design current (amp)
3.75	22.1	4.36	19.0	4.72	17.5

	Tilt at latitude-15					Tilt at latitude					Tilt at latitude+15				
	Corrected Load (AH/day)		Peak Sun (hrs/day)		Design current (amp)	Corrected Load (AH/day)		Peak Sun (hrs/day)		Design current (amp)	Corrected Load (AH/day)		Peak Sun (hrs/day)		Design current (amp)
Jan	82.7	/	4.01	=	20.6	82.7	/	4.56	=	18.1	82.7	/	4.84	=	17.1
Feb	82.7	/	4.65	=	17.8	82.7	/	5.01	=	16.5	82.7	/	5.08	=	16.3
Mar															
Apr															
May															
Jun															
Jul															
Aug															
Sep															
Oct															
Nov	82.7	/	4.29	=	19.3	82.7	/	4.90	=	16.9	82.7	/	5.22	=	15.8
Dec	82.7	/	3.75	=	22.1	82.7	/	4.36	=	19.0	82.7	/	4.72	=	17.5

Now, select the smallest design current and corresponding peak sun.

ie, Peak Sun = 4.72 hrs/day, and Design current = 17.5 amps. The corresponding 'tilt angle' is 55 deg.

The winter season load is expected to be the largest, and the system is sized to meet this load.

(A) To estimate the Battery system size:

The battery selected is ' Trojan' Lead Antimony battery with a 'nominal voltage' of 6 volts and 'Rated capacity' of 350 AH.

Corrected load (Amp-Hrs):		82.7
Storage days:		6
Max. depth of discharge:		70%
Derate for temperature:	1	
Required battery capacity	=	(82.7) x (6) / 0.7 / 1
	=	709 AH
Capacity of the selected battery:		350 AH
Therefore, Batteries in parallel	=	709 / 350
	=	2
Nominal System voltage	=	24 volts
Nominal battery voltage	=	6 volts
Therefore, Batteries in series	=	24/6 = 4
ie., Total no. of batteries:		2 x 4 = 8

Usable Battery capacity : (Batteries in parallel) x (capacity of the selected battery) x (max. depth of discharge) = 2 x 350 x 0.7 = 490 AH.

(B) To estimate System Array size:

The selected PV modules are Siemens M75 type with voltage, 15.9 volts and current, 3 amps at STC; Voc= 19.8 volts, Isc = 3.4 amps . Voltage at the highest expected temperature is 14.4 volts.(This information is furnished by the manufacturer)

Now,

Design current: 17.5 amps (from above)

Module derate factor: 0.9

Derated Design current: 17.5/0.9 = 20 amps

Rated module current: 3.0 amps

Therefore, Modules in parallel = 20/3 = 6
(rounded to conservative design)

From (A), Voltage required for Load = 1.20 x Nominal battery voltage x Batteries in series

= 1.20 x 6 x 4= 28.8 volts

From the manufacturer's data, Highest temp. module voltage = 14.4 volts

Therefore, Modules in series= 28.8 / 14.4 = 2.

Total no. of Modules = 2 x 6 = 12.

Array Rated current = Modules in parallel x Rated module current

= 6 x 3 = 18 amps

Array Short Circuit current = Modules in parallel x Module short circuit current

= 6 x 3.4 = 20.4 amps

Array Rated Voltage = Modules in series x Rated Module voltage

= 2 x 15.9= 31.8 volts

Array Open Circuit voltage = Modules in series x Module Open Circuit voltage

= 2 x 19.8 = 39.6 volts

The details of the Switches and Protection components used in the System are given in the table 3.

Circuit	Protection Device	Rated Current and Voltage
1.PV Array output	DC fused disconnect switch	30 amps/250 volts
2.PV Array output	DC fuse	30 amps/250 volts
3.Controller to Inverter	DC fused disconnect switch	250 amps/250 volts
4.Controller to Inverter	DC fuse	250 amps/250 volts
5.AC distribution	AC circuit breakers	15-20amps/220volts
6.Load input to AC load panel	AC manual switch	60 amps/220 volts
7.Battery charger	DC manual switch	100 amps/250 volts
8.Battery	DC in-line fuse	250 amps/250 volts

The designed System circuit with various components and protection devices is shown in the figure.

The methodology adopted here to estimate the array size and battery size are almost the same as discussed in the text.

Glossary

Air Mass (AM)	:	*It is the thickness of the atmosphere at sea level; for instance, the thickness of atmosphere is minimum at sea level when the sun is directly overhead and is called Air Mass 1 (AM 1).*
Ampere-Hour	:	*The battery capacity is referred in terms of 'Ampere-Hour'. An ampere-hour (AH) is equivalent to carrying one ampere current for one hour or 10 amperes for one-tenth hour.*
Array Mounting	:	*The mounting type used to fix solar PV array in a PV System. Rock mounting, direct mounting, standoff mounting and integral mounting are the types of mountings in use.*
Band (Energy) Gap	:	*It is a characteristic parameter of a semiconductor,- the energy required to separate an electron from the atom (the bonding) and making it a 'free' or conduction electron.*
BIPV	:	*Building Integrated PV is a PV System integrated into the building as a building element which produces electric power.*
Blocking diode	:	*A device fitted in the PV array to avoid damage to the modules due to the backward flow of electricity from the array due to any fault or array string mismatch.*
Bypass diode	:	*A device included to prevent hot-spot damage arising due to shade falling on the modules.*
Cascade/Tandem Cells	:	*Solar Cells made of materials with different band gaps are stacked together so that each cell converts a small range of solar spectrum close to its band gaps.*
Cd S	:	*It is Cadmium Sulphide used as a layer in p-n junction solar cells.*

Cd Te	:	*It is Cadmium Telluride,- a semiconducting compound popularly used in solar cell fabrication (CdTe Solar Cells).*
Charge Controller	:	*It is an electronic instrument which regulates the electric power from the solar module/array.*
CIS	:	*It is Copper Indium diselenide,- a semiconducting compound used in the fabrication of Solar Cells, (CIS Cells). Gallium is also added sometimes to make it CIGS.*
CSS	:	*It is close spaced sublimation,- a method popularly used in the fabrication of CdTe Solar Cells.*
CVD	:	*It is chemical vapor deposition,- another method to prepare thin films of materials, and there are different principles of operation in CVD.*
Deep-cycle Battery	:	*A battery used in solar PV Systems and is characterized by long life.*
Diffuse Radiation	:	*The solar radiation scattered by dust, water molecules and so on in the atmosphere and reaching the earth.*
Direct Radiation	:	*The radiation coming from the Sun directly.*
DoD	:	*It is the Depth of Discharge which decides the 'life' of a battery and most common DoDs are 10%, 20% and 50%.*
Global Radiation	:	*The sum of the direct and diffuse radiations is 'global' radiation.*
Grid-connected System	:	*It is one type of PV system that generates power, but connected to utility grid also.*
Hetero-Junction Cell	:	*Two different semiconducting materials form the p-n junction. Examples are CIS- and CdTe- based solar cells.*
Homo-Junction Cell	:	*Solar cell made of p- and n-types of the same semiconducting material. The crystalline silicon cells are of this type.*
Hybrid System	:	*It is a PV system which includes which has another source of energy system such as wind/diesel generator.*
Inverter	:	*It is an electronic device which converts the DC power from the solar module/array into AC power.*
MPPT Charge Controller	:	*A Maximum Power Point Tracking charge controller is a device which permits the PV array to operate at the voltage that produces maximum power output.*

n-semiconductor	:	*A semiconductor 'doped' with pentavalent atoms such as Phosphorous, to create 'extra' free electrons.*
One Sun	:	*It is equivalent to 1000 Watts per Square Meter.*
Open-circuit Voltage	:	*In an illuminated p-n junction, if the load is an open circuit, the corresponding voltage across the junction is open-circuit voltage.*
Photovoltaic Effect	:	*An effect in which when light falls on a p-n junction device, an electric current is produced across the junction.*
p-i-n Solar Cells	:	*The design used in the fabrication of amorphous silicon solar cells.*
p-semiconductor	:	*A semiconductor 'doped' with trivalent atoms such as Boron, to create 'extra' holes.*
Pyranometer	:	*An instrument to measure the 'global' and 'diffuse' solar radiation.*
Pyrheliometer	:	*An instrument to measure 'direct' solar radiation.*
Short-circuit Current	:	*If a p-n junction is illuminated and the terminals are shorted, the current that flows in the junction is short-circuit current.*
Silicon Wafer	:	*It is a thin sheet of silicon, a few microns thick, sliced from silicon ingot.*
Silicon	:	*It is a popular semiconducting material used in Solar Cell production. Single-crystal Silicon possesses perfect lattice order, Polycrystalline silicon contains grains randomly placed within which there is lattice order. Amorphous silicon has 'glassy' structure with no lattice perfection.*
Solar Array	:	*A set of solar modules connected in series/parallel to obtain desired electric power.*
Solar Cell	:	*It is an electronic device which converts solar energy directly into electrical energy.*
Solar Module (panel)	:	*An environmentally protected assembly of solar cells designed to generate required electric power at specified voltage and current rating.*
Solar Radiation	:	*It is the radiation coming from the sun and the total radiation at any point on earth is called 'Global Solar Radiation'.*
Sputtering	:	*It is a technique used to form thin layers or films of materials on substrates. There are different types of sputtering, such as DC and RF.*

Stand-alone System	:	*It is a PV system that generates electric power and stores in the battery, which is not connected to grid supply or a backup supply.*
Sun Simulator	:	*An instrument fitted with Xenon lamp or Quartz halogen lamp to give an intensity of 1000 Watts/Sq. meter with uniform distribution over the test bed.*
Thin Film Solar Cell	:	*A Solar Cell where in the p-n junction is formed by thin films/ layers of concerned materials.*
ZnO	:	*It is Zinc Oxide, used as a 'window' layer (transparent layer) to allow the solar radiation into the cells.*

Index

N

O

P

R

S

T

W